The Economic and Social Dynamics of Biotechnology

Economics of Science, Technology and Innovation

VOLUME 21

The titles published in this series are listed at the end of this volume.

The Economic and Social Dynamics of Biotechnology

Edited by

JOHN DE LA MOTHE
PRIME, University of Ottawa

And

JORGE NIOSI
CIRANO, University of Quebec at Montreal

KLUWER ACADEMIC PUBLISHERS
Boston / Dordrecht / London

Distributors for North, Central and South America:
Kluwer Academic Publishers
101 Philip Drive
Assinippi Park
Norwell, Massachusetts 02061 USA
Telephone (781) 871-6600
Fax (781) 871-6528
E-Mail <kluwer@wkap.com>

Distributors for all other countries:
Kluwer Academic Publishers Group
Distribution Centre
Post Office Box 322
3300 AH Dordrecht, THE NETHERLANDS
Telephone 31 78 6392 392
Fax 31 78 6546 474
E-Mail <services@wkap.nl>

Electronic Services <http://www.wkap.nl>

Library of Congress Cataloging-in-Publication Data

The economic and social dynamics of biotechnology / edited by John de la Mothe and Jorge Niosi.
p. cm.-- (Economics of science technology and innovation ; 21)
Includes bibliographical references and index.
ISBN 0-7923-7922-5 (alk. paper)
1. Biotechnology--Social aspects. 2. Biotechnology industries. I. de la Mothe, John. II. Niosi, Jorge. III. Series.

TP248.23 .E26 2000
338.4'76606--dc21 00-058420

Printed on acid-free paper
Printed in the United States of America

CONTENTS

PART IV: IMPACTS

PART V: CONCLUSION

CONTRIBUTORS

Anthony Arundel, MERIT, University of Limburg.

Bo Carlsson, Case Western Reserve University.

John de la Mothe, PRIME, Faculty of Administration, University of Ottawa.

G. Bruce Doern, CRUISE, Department of Public Administration, Carleton University and Department of Politics, Exeter University.

Gunnar Eliasson, Royal Institute of Technology, Stockholm.

Vincent Mangematin, Université Pierre Mendès.

Maureen McKelvey, TEMA, Linköping University.

Jorge Niosi, CIRANO, University of Quebec at Montreal

Bill Pattinson, Directorate of Science, Technology and Industry (DSTI), OECD.

Antoine Rose, Science, Innovation and Electronic Information Division, (SIEID) Statistics Canada.

Jacqueline Senker, SPRU, University of Sussex.

Nico Stehr, Gerhard-Mercator-Universität Duisburg.

Brigitte Van Beuzekom, DSTI, OECD.

Andrew Wyckoff, DSTI, OECD

Cherisa Yarkin, BioStar, University of California, Berkeley.

ACKNOWLEDGEMENTS

This book is the result of a happy coincidence. Well before this joint project was discussed or even conceived, both editors were each – in parallel – engaged in research looking at biotechnology, one (Niosi) with Statistics Canada and the other (de la Mothe) with CONACYT in Mexico and the RAND Corporation in Washington. But in making this volume a reality we have benefited from the input and assistance of a number of individuals and institutions. In February 2000, an Advanced Research Workshop, co-sponsored by Statistics Canada and the Program of Research on Innovation, Management and Economy (PRIME), was held in Ottawa, Canada. The participants included policy makers, biotechnology practitioners, as well as international scholars and experts. To all who were engaged in this conversation, thanks must be given. Thanks, too, must go to Kevin Lynch – then Deputy Minister of Industry Canada – whose interest and support made the meeting a success. For essential logistical and research support, high praise must be given to Tyler Chamberlin and Heather MacKinnon. Of course, Ranak Jasani has once again been a wonderful partner in working with us towards the production of this book, and Bo Carlsson and Cristiano Antonelli have once again invited us into their fine series. Thanks to all.

PREFACE

This is the third book from a project that examines technological change from economic, social and statistical perspectives. The previous two books dealt with innovation: the first with its regional aspects; the second, with the impacts of information and communication (ICT) technologies. The present volume continues this project with a focus on another key transformative technology - biotechnology.

Biotechnology and the ICTs are changing our lives, rapidly, and in different ways. While ICTs change the way we live, work, educate and entertain ourselves, biotechnologies have a direct effect on our health and environment and on the plant and animal food that we eat. The associated risk and safety issues ensure a different regulatory environment for the introduction of new medicines or foods from that affecting, for example, the producer of a new software product. However, in both cases new or significantly improved products or processes are being introduced to the market and this is innovation.

The activity of biotechnology involves research and development, invention and related intellectual property regulation, innovation, diffusion of the technologies throughout the economy and the society, and the human resources that are needed to make all of these things happen. The development of indicators of such activities, as well as their linkages, and their outcomes, is part of an on-going research program at Statistics Canada, which aims to provide a coherent picture of the science and technology system in Canada.

The discussion of the use and planned use of biotechnologies requires concepts and definitions which apply not just in one country, but in many, if there are to be meaningful comparisons of statistical measures and indicators of activity. The Organization for Economic Co-operation and Development (OECD) is in the process of reviewing these very questions with a view of developing guidelines for data gathering, reporting, and analysis. However, internationally comparable statistical information on the wide range of biotechnology activities is a long way from the work on R&D which has a history of at least four decades at the OECD. The combination of rapid change in the technologies, and their application, and the growing social concern about their impacts, suggests an immediate need for comprehensive and comparable indicators to support public policy debates on regulation, production, and use, of the products of biotechnologies.

The need to advance a research agenda with relevance to official statisticians, to policy makers, and to academic analysts was the motivation for a research workshop convened jointly by the Program of Research in Innovation Management and Economy (PRIME) at the University of Ottawa and by Statistics Canada in February 2000. Most of the papers in this book are derived from that workshop, while others have been invited to complement the collection.

The stage is set in the Introduction and the section on Frameworks provides both tools and analytical insights which inform the discussion developed in later sections. The analytical tools are:

- a systems framework applied to analyse the economic and social dynamics of biotechnology;
- a competence bloc framework applied to analyse the industrial potential for biotechnology; and,
- an analysis of economic, political and social constraints which inhibit the development of products for which there is a social need.

The section on Measurement presents a body of statistical information on the activity of biotechnology and on firms that are principally engaged in biotechnology. It spans the aggregate statistical approach of national statistical agencies and that of case studies applied to two specific questions: the reasons for the high growth rates of some firms engaged in biotechnology; and, the impact of the University of California on firms engaged in biotechnology in the State of California. The section shows what can be done, and, by implication, what needs to be done to provide a comprehensive set of statistical indicators to support public policy debate.

In the section on Impacts, a bridge is provided from the previous sections by an overview of plans for measurement and analysis of biotechnology activities now underway at the OECD. This leads to a paper which examines the international regime for biotechnology, including trade, intellectual property, and safety. A national perspective is developed by papers on biotechnology in Australian and in French industries and the perspective shifts to the social and the policy arenas in the last two papers.

The Conclusion suggests some research directions for statisticians, for policy analysts, and for academics.

Fred Gault

Director, Science, Innovation and Electronic Information Division
Statistics Canada

PART I

INTRODUCTION

Chapter 1

TOOLS FOR ANALYSING BIOTECHNOLOGY

John de la Mothe
PRIME, University of Ottawa

and

Jorge Niosi
CIRANO, University of Quebec at Montreal

The analysis of innovation presents important and unique challenges. In the world of biotechnology these are even more pronounced. Intangibility, pervasiveness, trans-disciplinarity, multiplicity of sources and multiplicity of application areas all combine to make the analysis of biotechnology significant. And not only are the dynamics of biotechnology at variance with those of information and communication technologies (ICTs), but public concerns over BSE, GMOs, cloning and gene therapy – or what J.B.S. Haldane (in a Blakean way) referred to as 'perversions' – make biologically derived techniques and technologies of particular interest. However, the objective of this research volume is not to proclaim on issues of ethics or jurisprudence. This is, after all, - as Lord Dahrendorf has said – the Age of Schumpeter.

To analyze biotechnology in terms of its economic and social dynamics requires a deliberate set of tools that this book attempts to provide. A first step is to present a package of frameworks. Next we need to turn to aggregate statistics and then to case studies. Given that technological change and innovation have both local and global characteristics, attending to the international context provides a valuable tableau against which to assess dynamics. In this case, we look to comparisons as well as to institutions, which together establish the rules of the game, the international regimes within which we must operate. This is amplified through country and regional studies, and this is followed by a consideration of impacts at the social and policy levels.

FRAMEWORKS

Three frameworks which are related and which are deployed in this book are innovation systems, technological systems and competence blocs.

The innovation system perspective is useful in that it allows the consideration of numerous factors – including science, technology, finance, production, and distribution. It focuses attention on the strength of the interactions between players and in so doing deals with the dynamics of the system.

The technological system approach works well below the level of the sector and emphasizes the importance of technological evolution. By looking at complementarities and synergies, analysts can begin to examine the root causes of clustering in economic analysis. Moreover, this approach deals centrally with networks of knowledge and competence. With entrepreneurial spirits present, as well as with sufficient critical mass, synergistic clusters of firms, technologies and talents can be transformed into development blocs.

This brings us to competence blocs in which the demand side (unlike traditional science and technology policy) is key. Here, customers set the criteria for economic selection. Through competence blocs, analysts can deal with learning in the marketplace and bolster the network dynamics set out be evolutionary, innovation and technology-based frameworks.

DATA AND MEASUREMENT

In order to assess the performance of an innovation or technological system, data are required. Hence a number of authors here attend to questions of data and measurement. To anyone even subsidiarily aware of the knowledge-based economy and the problems of describing it in the System of National Accounts (SNA), the problem and importance of measurement is clear. This becomes particularly important in the realm of biotechnology as biotechnology has current and potential applications in a large number of industries, including pharmaceuticals, chemicals, mining, forestry, fisheries, agriculture and food processing. It is for this reason that biotechnology is considered to be pervasive. Aggregate statistics – drawing on surveys involving numerous firms and respondents – allow us a window on a variety of characteristics, including employment. Moreover, good aggregate statistics allows the evaluation of biotechnology's economic impacts. These include comparative data on the types of biotechnology that are in use, the distribution of firms using biotechnology, the application of specific types of biotechnology, and the effect of alternative technology on biotechnology applications.

Case studies illuminate the reasons why some biotechnology firms grow more rapidly than others, and how a strong university base can help in the generation and distribution of value and benefits.

Measurement is a local or national activity. But systems of innovation involve flows of knowledge, technology, information and talent that are international. The OECD provides the means of comparing data from 29 member countries. The international institutional context is also key to the function of firms engaged in biotechnology and a discussion of international regimes provides us with a global framework within which to see the comparisons developed by the OECD.

IMPACTS

Evaluating a technology, of course, is not simply a matter for measurers. Impacts and the flow of values are critical if a responsive governance structure is to be achieved. It is therefore essential that this volume considers shifts in the nature of the marketplace, of intellectual property, public policies, and the equitable distribution of benefits in an area that front-loaded by significant investment and development costs as well as by considerable discovery and testing costs.

CONCLUSION

Taken together, we learn much in this volume about the structure, institutions and architecture of the innovation process. We learn about the dynamic that is innovation, through which so much new value and creative destruction flows. Knowledge, learning, and information are centrally posed as the new currency. And perhaps most importantly, through our attention to biotechnology, we learn again that dynamic frameworks are requisite, that better data sets are essential, and adaptive public policies are needed. This does not remove us from the responsible need for transparency, public access, precautionary risk assessment, and so on, but it does allow us to see something clearly about the economic and social dynamics of biotechnology.

PART II

FRAMEWORKS

Chapter 2

A SYSTEMS FRAMEWORK FOR THE STUDY OF ECONOMIC AND SOCIAL DYNAMICS OF BIOTECHNOLOGY

Bo Carlsson
Case Western Reserve University

INTRODUCTION[i]

Over the last few decades there has been a growing recognition of the need for a systemic approach to the study of economic growth and technological change.[ii] Various explicitly or implicitly systemic interpretations have been offered in the literature, including industrial districts (Marshall 1890), development blocs (Dahmén 1950, 1989), technological systems (Hughes 1987), techno-economic paradigms/systems (Freeman, Perez 1988), and techno-economic networks.

Perhaps the most widely used concept in recent years has been that of *national innovation systems* (NIS) (Freeman 1988; Lundvall 1988 and 1992; and Nelson 1988 and 1993; and subsequently many others). This delimitation of the system has the advantage of focusing attention on the major institutional determinants of innovative activities. It reflects the concerns in the 1970s and 1980s about the competitive strength of national economies (mainly the U.S. economy and the West European economies) challenged by the unprecedented success of Japan. Porter (1990) has developed a similar approach.

However, the NIS approach has certain drawbacks. It tends to neglect the fact that (i) innovation systems differ greatly in character across industries and sectors; and that (ii) they are increasingly international. It also tends to neglect the regional dimension of such systems. Thus recently there has been a partial shift of focus, mainly in the direction of *technological systems* (TS) (Carlsson and Stankiewicz 1995, Carlsson 1995 and 1997), *sectoral systems of innovation* (Malerba and Orsenigo 1990, 1993, 1995; Breschi and Malerba 1997), and *regional innovation clusters* (Porter 1990; Saxenian 1994).

The different versions of the systemic approach to innovation each have their own strengths and weaknesses. On the whole they are complementary and, to a large extent, overlapping. The choice between the approaches reflects the type of issues one wishes to study. If one's main concern is the impact of institutional structures and the role of national policy, the NIS approach may be the appropriate choice. If the main concern is to understand the impact of technological change on economic growth, the TS approach may be most appropriate. The latter is in many ways similar to the sectoral approach and can often be combined with it.

The concept of "sectoral innovation systems" is based on the idea that different sectors or industries operate under different *technological regimes* which are characterized by particular combinations of opportunity and appropriability conditions, degrees of cumulativeness of technological knowledge, characteristics of the relevant knowledge base, and so forth. The approach results in much more finely grained analyses of the innovation processes and their institutional and organizational determinants than in NIS. It explains many striking differences in the industrial structure and dynamics of different branches. However, here too there are some problems.

The main shortcomings of the SIS approach are (i) that its concept of technological regime is largely static and unexplained, and (ii) that it takes as its point of departure well defined industries/branches. It is therefore less convenient when one tries to analyze the emergence of new systems of innovation or radical transformation of existing ones such as is currently happening in biotechnology.

The technological systems approach goes beyond the sectoral approach by emphasizing the importance of technological evolution as the factor behind the emergence of new technological regimes and the associated systems of innovation. Technological complementarities and synergies are viewed as the major factors leading to clustering of economic activity. In the past, such clustering was often based on the proximity to natural resources and markets. The resulting competence accumulation stabilized and strengthened the competitive advantages of the local "development blocs." Today, that clustering is increasingly a result of technological complementarities and based on knowledge flows rather than material flows. The concept of "technological systems" has been designed to capture that dynamic.

THE TECHNOLOGICAL SYSTEM FRAMEWORK

The technological system framework was defined originally as a network of agents interacting in a specific area of technology under a particular *institutional infrastructure* or set of infrastructures and involved in the generation, diffusion and utilization of technology. Technological systems are defined in terms of knowledge/competence flows rather than flows of ordinary goods and services. They consist of dynamic *knowledge and competence networks.* In the presence of an entrepreneur and a sufficient critical mass, such networks can be transformed into *development blocs*, i.e., synergistic clusters of firms and technologies within an industry or a group of industries (Carlsson and Stankiewicz 1995, p. 49).

This definition has been applied to a variety of technological and industrial fields including factory automation, pharmaceuticals, biotech industry, electronics and computer technologies, and powder technologies (Carlsson 1995 and 1997). In the course of these studies we have encountered a variety of conceptual and methodological problems. They arise from the fact that our approach strives to integrate three distinct dimensions: (i) the *cognitive structure* of technology; (ii) the *organizational* network of actors; and (iii) the *economic* analysis of the resulting development blocs. Each of these levels has a dynamic that interacts and influences those of the others, but is not reducible to them.

Clearly, the emphasis on technology is the defining characteristic of our approach. But technology does not, by itself, constitute a technological system. The latter requires a network of actors who, as it were, embody the system. Furthermore, we must be able to analyze that network as an economic organization operating in the context of the market economy. Let us therefore briefly discuss each of these dimensions of analysis and their interrelations.

Technology - The Cognitive Dimension

The concept of "technology" is notoriously ambiguous. This is not the place to consider all its aspects. For our purposes, the term connotes the sum total of intellectual resources necessary for the production of goods and services. Technology is knowledge, but it is a particular species of knowledge which is distinct from others, such as science. As a cognitive system technology has a structure, albeit one that continues to be poorly understood.

At a rather rudimentary level we can distinguish between two modes of technological progress. On one hand, technology can be seen as consisting of interrelated sets of competencies linked to an evolutionary trajectory of a particular artifact, such as, for instance, the aircraft. On the other, technology can be seen as a broader capability giving rise to and sustaining the development of multiple artifacts.

Needless to say, these distinctions are always a matter of degree and are sometimes hard to make in practice. Nevertheless, they are significant. Adopting one or the other view will strongly affect the focus of analysis. There is a logical connection between the "broad capability" approach to technology and the systemic approach to innovation.

But how is one to conceptualize that "broad capability?" There exists in the literature a number of different but closely related concepts seeking to do so. The best known are technological trajectories or paradigms (Nelson & Winter 1977, 1982; Dosi 1982) and technological regimes (Dosi 1982). Especially the notion of technological paradigm has been much discussed and theoretically elaborated. Unfortunately, from our point of view, it has two drawbacks: (i) it operates on an extremely high level of aggregation; and (ii) it is burdened by a number of conceptual difficulties inherited from its Kuhnian prototype.[3] In our earlier work we relied instead on the concept of "generic technology". However, that concept does not work very well either as the basis of the definition of innovation systems, since the latter always arise from a *combination* of many technologies, both generic and specific. Indeed, it is useful to think about technology as a combinatorial "design space" generated by a whole range of existing capabilities. Design spaces are formed by clusters of technical capabilities. Complementarities among these capabilities result in the clustering of technological opportunities in certain regions of that space. It is that clustering which, under appropriate economic and institutional conditions, gives rise to technological systems.

Of course, the emergence of new technological clusters or the transformation of existing ones can be initiated by a particular technical break-through. In the case of "new" biotechnology, the triggering events were the discovery of recombinant DNA and monoclonal antibodies techniques. However, it is more common that several parallel developments are necessary. In either case, major economic impact can occur only when these novelties are integrated within a wider cluster of competencies, both old and new.

Actor Networks and Institutional Infrastructures – The Organizational Dimension

Technological systems consist of networks of highly diverse actors. These form "technological communities" composed of all individuals (practitioners, researchers, mediators) who contribute to growth and structuring of the technology in question. These individuals are spread across a variety of organizational environments: companies, universities, other R&D and educational organizations, public bureaucracies, industry organizations, and so forth. They include inventors, researchers, engineers, entrepreneurs, managers and bureaucrats.

Technological systems are heterogeneous in that they involve both market and non-market interaction in three types of network: buyer-supplier (input/output) relationships, problem-solving networks, and informal networks. While there may be considerable overlap between these networks, it is the problem-solving network which really defines the core of the system: What are the sources of knowledge? How do complementarities arise? And where do various actors in the system turn for help in solving technical problems?

Buyer-supplier linkages may be important, the more so the more technical information is transmitted along with the transactions and the less so, the more commodity-like the transactions are. Sometimes the most important technical information comes from sources (e.g. universities and research institutes) separate from buyers and sellers. This is certainly true in science-based fields such as biotechnology. Sometimes the informal, mostly personal, networks established through professional conferences, meetings, publications, etc., are important channels of information gathering and sharing.

The ability of the technological communities effectively to promote technological progress and economic growth is largely a function of the incentive systems as well as institutional and organizational arrangements under which they operate. These conditions determine communication patterns in the systems, the R&D agendas of different actors, and the modes of knowledge accumulation. They are the results of both deliberate efforts of science and technology policy makers and of spontaneous organizational evolution.

The composition and structure of the actor networks are closely related to the cognitive characteristics and dynamics of the technologies involved. If the technological growth is to proceed rapidly, the two must become correlated. However, that correlation may either fail to materialize, or it occurs only with a significant time lag. Organizational and institutional inertia are the major sources of such lags.

Development Blocs – The Economic Dimension

Technological systems are embedded in economic systems. The actors within the technological systems are either economic actors or operate within or through such actors – firms and other private as well as public organizations. The economic organization impacts the technological system in three ways: (i) it shapes the relationships within the actor network; (ii) it determines the allocation of resources to knowledge creation and other innovative activities; and (iii) it determines the degree of exploitation of the opportunities generated within the system.

From the outset we have focused on the notion of "development blocs"(Dahmén) as the tool for the analysis of economic dynamics of technological systems. Effective innovation and economic growth depend on the ability of economic actors to exploit emerging resource complementarities, in our case - technological ones. Development blocs are synergistic networks of agents (primarily firms) exploiting such complementarities.

It is possible to identify a number of conditions under which development blocs can emerge and develop. Factors such as proximity, critical mass, entrepreneurial drive, and many others play important roles. Eliasson & Eliasson (1997) have developed the concept of "competence bloc" which describes the composition of development blocs in terms of distinct techno-economic roles and competencies. This is discussed in the following chapter. These typically include competent customers, innovators, entrepreneurs, venture capitalists, and industrialists. The innovative success depends on the completeness of the bloc; if one or more of these elements are missing, e.g., because of lack of incentives, the competence bloc never develops fully.

It must be observed that there is no one-to-one relationship between technological systems and development or competence blocs. A technological system may give rise to or support several development blocs, which of course may be interrelated. The large, comprehensive sectors of the economy, such as health care, communications or energy, are likely to consist of several interrelated development blocs, but this does not imply that the sector has a single unified technological system.

While the general structure of development/competence blocs may be similar across many technological systems, it is clear that there also are large differences across various fields of technology and over time. Thus the exact composition of the required role set will depend both on the general logic of economic activity and on the specific nature of the actor networks forming the technological system which in turn reflects the cognitive dynamics of the technologies involved. In other words, just as there is a correlation between the cognitive structure of technology and the actor networks, so there is a correlation between the former two and the development/competence blocs. However, here again the correlation cannot be expected to be perfect or stable over time.

Dynamic Character of Technological Systems

To recapitulate: from the cognitive point of view, technological systems emerge and evolve around complementary clusters of technological capabilities that jointly create new economic opportunities. Such clusters tend to be highly dynamic. Their composition changes over time, sometimes quite rapidly. The boundaries of the clusters are therefore rarely clear. In fact, the contemporary technology is characterized by frequent "technological fusions" and increasing technological commonalities among earlier technologically unrelated activities.

Technological complementarities create powerful incentives for interaction and for the creation of organizational and economic networks. The pattern of interaction and networking becomes progressively stabilized and institutionalized in terms of products and production systems, firms, industries, engineering disciplines, and infrastructures. In short, they are gradually transformed into full-fledged technological innovation systems.

Generally one would assume that the fit between the three components of the technological system discussed above should improve over time and be very strong in mature systems. But this final state arises only rarely and when it does, tends to be of limited interest. It arises rarely because most technological systems, especially contemporary ones, are either displaced or radically transformed long before they reach maturity. It has limited interest because maturity generally means that the system cannot evolve any further, that it has reached a dead end.

From this it follows that it is the dynamic characteristics of technological systems rather than the degree to which they fit some image of an ideal "final state" that should be our main concern. The dynamic is created by the tension between the logic of technology on the one hand and the extant social structures (actor networks and development/competence blocs) and incentives on the other. Systems that are capable of resolving that tension prosper, while those that are not fail. But neither the failure nor success need be permanent. Further development of technology may easily reverse the tables.

The analysis of the highly dynamic and structurally loose systems presents a considerable challenge. Let us therefore consider some methodological issues that arise in the course of such work.

METHODOLOGICAL ISSUES

Three major methodological issues arise when one attempts to study technological systems empirically. First, there is the question of how best to capture and describe the *content* of such systems in terms of the three dimensions of analysis (cognitive, organizational, and economic). Secondly, what is the best way to *delimit* the specific objects of analysis? Thirdly, how can system performance be evaluated?

Contemporary technological systems are very large and complex. Geographically they are widely dispersed, often global. Due to frequent technological fusions, their technological scope is great and growing. The nature of this problem is well illustrated by the recent convergence between computing and communication technologies. Under such circumstances it is difficult to analyze technological systems in their entirety. We are forced to focus our attention on selected parts of the system, hoping that these parts are sufficiently representative of the whole.

The selected parts of the system can be delimited using such criteria as geographic area, economic sector, industry, or a product area. One can also attempt to limit the scope of investigation by focusing on less inclusive clusters of technologies. Out choice of criteria will reflect our particular research interests and practical constraints. But we must be aware that every choice we make carries with it a price in terms of the comprehensiveness of the analysis. Let us now discuss in some detail each of the methodological issues listed above.

Identifying Technological Clusters

The empirical delimitation of technological areas (clusters) is extremely difficult, especially in the early stages of the development of the cluster. Often such delimitation can be done only indirectly and approximately. There are several reasons for that:

The evolution of technological clusters is inherently unpredictable, in the sense that all evolutionary phenomena are inherently unpredictable. Hence they can be accurately described only *ex post*.

Technologies are largely intangible and therefore not directly observable. The existing artifacts offer us only glimpses of parts of the underlying design space. Scientific journal articles and patents, too, are only proxies.

New technological clusters and systems are always embedded in a historically given institutional and economic matrix and can be difficult to disentangle from it.

There are no sharp boundaries between the different regions in the total web of technologies. In fact, it seems that that web (or "design space) is becoming increasingly unified. Hence the cluster boundaries are always permeable, relative, and changeable.

We are interested in identifying emergent technological clusters which give rise to new economic opportunities. Generally speaking, one can identify such clusters using one or both of the following methods: (i) exploiting *expert judgment*; and (ii) carrying out *bibliometric and technometric* studies. In each case we seek to assess the "proximity" and/or coherence of a given set of technical capabilities.

Expert Judgment

To be able to delineate the system, we need to understand what are the composition, coherence, and boundaries of the emerging technological clusters. Expert interviews, analysis of professional literature and of programmatic statements of various kinds, as well as surveys of technological foresight studies may yield good approximate definitions of important technological clusters. The great advantage of the approach is that it is easy to use and that it allows one to capture emergent phenomena which elude more quantitative approaches. The disadvantage its inevitable subjectivity – which, however, can be mitigated by using a sufficiently wide set of sources.

Biblio- and Techno-metrics

Various formal and quantitative approaches have been developed to measure technological proximity/distance in terms of the linkages between the knowledge fields involved (Granstrand and Jacobsson 1991, Ehrnberg and Sjöberg 1995)[4]. Bibliometric analyses (citation, co-citation, co-classification, and co-authorship) of journal articles and patents often give interesting insights into the structure of technological clusters. This kind of data can be used jointly with other measures of proximity such as, for example, expert judgment (Grupp 1996) or the data on the *amount of retraining* that engineers specialized in one of these fields need in order to be able to make a contribution to the other.

Generally, these document-based methods work better with journal literature than with patents – both as regards the density of citation networks and the precision of the coding classes. Both methodologies (expert judgment and bibliometrics) share the weakness of not providing any clear cut-off criteria, i.e., definition of cluster boundaries. These difficulties are compounded by the fact that technological systems are not static but continually evolve with changing content of technologies and products. Over time, *new* sub-technologies may emerge which need to be included in the system. Even more importantly, new *complementary technologies* may develop, transforming the entire cluster. Due to this dynamic character of the underlying competence base, the boundaries of a technological system may need to be broadened (or in other cases narrowed), perhaps leading to a change in the set of actors (components), relationships, and attributes to be included. For instance, in biomaterials there has been a shift of emphasis from synthetic to biological materials, and the delineation of the field has changed due to the introduction of new competence within, e.g., biotechnology (Rickne 2000). In this case, the relation between sub-technologies has shifted, resulting in links between synthetic and biological materials.

Hence, in longitudinal or historical studies we may need to redefine the boundaries of the system as it evolves. There is, therefore, no unique and always valid way of delineating a technological system. This does not, of course, make the empirical delineation any simpler.

Identifying the Actors

There are basically three approaches that can be used to identify the relevant population of actors: (i) the use of proxy populations; (ii) document-based analysis; and (iii) surveys.

Proxy Population

If we are prepared to take the membership in some well defined industry or sector as a proxy for membership in a given technological system, then the identification of actors, at least at firm level, is comparatively straightforward. We can simply rely on statistical data such as input-output tables and production and trade statistics.[5] There are also industry associations and other organizations that have an interest in cataloging firms in a specific product area. However, care must be taken in comparative studies where industry associations in different countries may have different degrees of success in organizing the industry and may set different industry boundaries.

This method will not be very satisfactory in the fields where technological development is very rapid and the established industrial classifications are quickly becoming outdated. Furthermore, the type of industrial data discussed here are not likely to identify non-industrial actors, such as public R&D institutions and universities.

Document-Based Analysis

Bibliometric and technometric analysis can be used both to identify technological clusters and the actors associated with those clusters. The methodological weaknesses and strengths are similar in both cases. These methods are generally good at identifying actors in the research system (particularly academia) but less effective in identifying firms and/or individuals within firms.

Mapping the competence base of firms is commonly done with by using patents (e.g. Miyazaki 1994, Jacobsson et. al. 1995, Praest 1998). However, there are at least three problems involved in using patents, apart from those conventionally listed (Pavitt 1988)[6].

First, a general problem with patent-based methods to identify a population on knowledge-based criteria is that the patent classification system is not necessarily structured around specific knowledge areas. Quite frequently, the classes are functionally based.

Second, patent holding does not necessarily reflect a deep knowledge in a particular knowledge field.

Third, patents reflecting knowledge to develop a particular technology, for instance microwave technology, may be found in many classes and a quite elaborate method may need to be devised to identify these (see, for instance, Holmén and Jacobsson 1998).

Survey of Actors

This approach usually involves some form of "snowballing sample." Starting with some initial population we can ask questions about additional members. Ideally the process continues until one arrives at a point where additional inquiries yield no significant additions to the population. However, the result may critically depend on the choice of the initial population. Furthermore, the method assumes that the firms or other actors are aware of at least some other actors who master the specific technology area. This may not always be true. Another problem is that the number of actors may expand and exceed the practical limitations of the study. Two other weaknesses of the approach are (i) its cumbersome nature; and (ii) the risk that one misses the emergent developments, i.e., ones which have not yet generated large and stable "technological communities".

Of course, given the inherent uncertainties in each method, it may be useful to use them in combination. For example, Rickne (2000) combined three methods. Biomaterials technology, which she has studied, can be incorporated into many products. The first step was to identify these products and consult industry associations and directories for firms producing them. Second, interviews with these firms and associations pointed to further actors (researchers, firms, and organizations) which in turn were contacted (snowball). Third, citations of important inventions verified and broadened the set of actors. Also Holmén and Jacobsson (1998) supplemented the snowball method with a patent-based method in order to reduce the risk that the population was not fully identified. Indeed, in the patent-based method, they identified a few actors that the snowball method had missed.

Identifying Development/Competence Blocs

The technological opportunities created by a technological system will not be effectively exploited unless they are supported by appropriate development/competence blocs. These consist of competent customers, innovators, entrepreneurs, venture capitalists, mechanisms of exit such as initial public offering (IPO) markets, and firms capable of exploiting the technology at sufficient scale (Eliasson & Eliasson 1997). In order for sustainable economic activity to get under way and succeed in the market, all these functions must be in place.

It is far from obvious precisely what constitutes the relevant development/competence bloc in a given instance. A technological system can give rise to multiple competence blocs and a single competence bloc may draw on several relatively independent technological systems. Transformations at the level of technological systems will cause corresponding transformations at the level of development blocs, i.e., new blocs are created and old ones disintegrate or are recombined.

Clearly a development/competence bloc cannot be identified with a single industry since it is organized around vertical (user-supplier) relationships as well as horizontal (competitive) ones. Or, to put it differently, it is organized around complementary as well as shared technological and economic competencies. In order to identify development blocs it may be useful to pursue a two-pronged strategy. One may start by analyzing the pattern of economic and technological interdependencies within a broad economic sector (consisting of several interrelated development blocs) and, at the next stage, try to separate the individual blocs within the sector.

Selection and Delimitation of the Objects of Analysis

The phenomenon of technological convergence and integration which characterizes many areas of today's economy prompts one to define technological system in an inclusive open-ended fashion. Unless we do so, some of the most important features of the innovation processes may be lost. However, the inclusive approach poses big practical problems. The systems it identifies may be too large and too complex to be studied empirically in their entirety. We are then forced to select some parts or aspects of them and analyze those in the hope that they are representative enough to shed useful light on the whole. What are then the principal ways of subdividing the systems into meaningful subsystems?

There are two main ways to perform such sub-division: (i) spatially; or (ii) by level of analysis. Precisely how they are used will depend on both theoretical and practical considerations. In either case it is crucial that the criteria used result in a clear-cut definition of the populations/objects to be empirically studied.

Delimiting Spatial/Geographic Units

Technological systems, particularly those in high technologies, tend to be international in character. However, the intensity of interactions in such systems varies geographically. There are two types of natural geographical sub-units here: countries and regions. The choice of country as a sub-unit is advantageous if our aim is to analyze the impact of broad policy and of institutional and cultural factors on the innovation processes. On the other hand, if we aim at the understanding of the dynamics of technological systems and development blocs at the micro level, a regional focus may be preferable. Indeed, most high technologies exhibit a strong tendency towards regional clustering, suggesting that development and competence blocs may to a considerable extent be local.

Delimiting the Levels of Analysis

There are basically four approaches available: (i) selection of sub-clusters of technologies, (ii) selection of a product (area) or industry; (iii) selection of a development bloc; or (iv) selection of an economic sector. Obviously we can combine these approaches in various ways.

Let us *first* consider the selection of sub-clusters. For example, within the technological system of telecommunications we may decide to focus on a particular technological sub-cluster such as, for example: digital signal processing. That technology may be used in a number of different products (e.g., mobile phones, control systems, etc.). Using this as a criterion we will analyze only those products or development blocs which include digital signal processing. Here the products are not the main focus of analysis. Our main concerns are the relations between technologies and the diffusion of technologies into different applications. For instance, Holmén (1998) studied microwave antenna technology that is incorporated into many highly diverse products including mobile phones, microwave ovens, military radar and automatic doors. Through those applications we can identify the relevant customer population.

In the *second* approach we take a product as the initial seed from which the system is defined. For example, an industrial robot consists of a number of technologies, e.g., drive, sensor and control technologies. But the technologies are not the primary criteria for delimiting the system. Instead, it is the artifact which is studied, and we want to study the links to its customers (in this case customer groups C1 and C2). Generally speaking, this approach to delimiting the object of investigation is fairly straightforward. The main problem here is how far we can generalize from such product-centered clusters to the larger technological system to which they belong. Clearly this is a matter of sampling.

The *third* approach is more inclusive. Here we aim at identifying distinct development or competence blocs. Our point of departure will typically be a chain of linked products rather than an individual product. A case in point would be the forest cluster in Finland which consists of key products such as paper and pulp and upstream and downstream industries such as paper machines and printing plants (Ylä-Anttila 1994). Another example is our earlier work on factory automation (Carlsson 1995). The approach is fraught with difficulties: there is always the issue of where the boundary of the system lies. There is no reason to hide that the delineation may often be somewhat arbitrary and partly based on informed guesses by the researcher (Porter 1998).[7]

These complexities increase greatly when we move to *the fourth* approach in which we use the sector as the criterion. Here we define the technological system in relation to a coherent set of interrelated economic activities such as the telecommunication system, health care system, or energy production and distribution system. It goes without saying that this delimitation results in an extremely complex object of analysis. But it is justifiable when the industries and development blocs within the sector are undergoing technological convergence, as is the case in the health care system today. The inherent risk connected with this approach is, however, that the underlying technological dynamics will be too difficult to analyze and therefore set aside. When that happens, we risk sliding back into the analysis of sectoral innovation systems rather than technological systems.

Evaluating the Performance of the System

The study of technological systems inevitably leads to three sorts of questions: (i) how do the dynamics of different technological systems compare with each other? For instance: is the technological system of biotechnology today essentially the same as, say, that of microelectronics two decades ago? (ii) how well developed or mature is a given technological system at a given time? Are there some easily identifiable indicators of maturity which apply to all technological systems? Finally, (iii) how successful are different economic systems (national or regional) in supporting the rapid evolution of these systems and in exploiting their economic potential?

The answers to these questions depend on our ability to define and measure the performance of technological systems and associated development blocs. Given the complexity of technological systems and their dynamic character, the task is daunting. How would one, for example, compare the performance of two economic regions as regards the technological system of telecommunications and computing? Does the fact that one of the regions performs poorly as regards the manufacturing of computing and telecom equipment mean that its technological system is weak? What if this very region is the main supplier of key scientific and technological inputs to industry located elsewhere? There are two methodological issues here; (i) is the particular choice of the spatial/administrative units meaningful? And (ii) is it sensible to use any global performance indicators for systems which are so heterogeneous and loosely structured? Should one not instead aim at a whole battery of partial indicators each gauging a different dimension of the technological system?

The answer depends to a large extent on how the objects of study have been delimited. In our early work on factory automation (Carlsson 1995), the main performance indicator was the extent of *diffusion* of factory automation in Sweden, as compared to other countries, using conventional diffusion analysis.[8] Keeping the product, or industry, as the unit of analysis we could, in later studies (Carlsson 1997), use patents to calculate the *revealed technological comparative advantage* in, say, electronics, as an indicator of the *generation* of knowledge and conventional performance indicators of the *use* of technology, such as *market shares and exports.* These types of indicators can also be used when the unit of analysis is a competence bloc. Hence, satisfactory ways of measuring the performance in terms of the generation, diffusion and use of technology, in the sense of an artifact, are available at least for well-developed systems.

Thus, in the case of factory automation, we could use them with relative success because the technological system under consideration was sufficiently mature, its development bloc well crystallized and represented within a single country. This is by no means always the case. Granberg's study of powder technologies suggested that the development blocs involved were international in character so that the use of an aggregate down-stream performance evaluation at the national level would distort the true strength of Sweden in this field of technology (Granberg 1997). The same study also indicates that it is not possible to use the same measurement yardsticks when technological systems (or their subsystems) differ in maturity. Powder metallurgy is a far more mature area than powder ceramics. In this case, the use of aggregate downstream indicators (such as sales, exports, or profits) may lead one to a logically inadmissible conclusion that Sweden has been more successful in the former than in the latter. To make such judgment we would first have to "normalize" both sets of data using some sort of technology maturity model. Unfortunately, we do not have such model. This is a serious issue particularly whenever one deals with large inclusive technological systems or delimits one's object of analysis sectorally. In such cases one is bound to deal with several sub-technologies and even distinct development blocs which find themselves at very different stages of development.

Clearly, whenever we wish to carry out comparative analyses or evaluate the success of a technological innovation system we need to ensure that the objects of analysis are sufficiently comparable. Furthermore, we need to make sure that all important elements of a technological system are monitored and assessed. Rickne (2000) identifies three types of performance measures for assessing the performance of technological systems. The relative weight attached to these indicators will vary depending on the system's maturity.

Figure 2. Examples of Performance Measures for an Emerging Technological System

Indicators of generation of knowledge	**Indicator of the diffusion of knowledge**	**Indicators of the use of knowledge**
Number of patents Number of engineers or scientists Mobility of professionals Technological diversity, e.g. number of technological fields	Timing /The stage of development. Regulatory acceptance Number of partners/ Number of distribution licenses	Sales Employment Growth Financial assets

Source*:* Rickne (2000)

The ability of the system to generate knowledge is assessed using four indicators. The first is the conventional patent indicator, revealing the volume and direction of the technological capabilities in the system. A related, and second, indicator is the number of scientists and/or engineers active in the technological fields. Not only the volume of activities matters but also cross-fertilization of different technologies, ending up in new and difficult-to-foresee combinations of knowledge. Here, the mobility of professionals, with a subsequent diffusion of their knowledge into new technological fields, may be a performance indicator (Rappa, 1994).

The fourth indicator is even less conventional. There is often a large uncertainty regarding which of a whole range of technological approaches will succeed in reaching the market in an immature system. This is particularly so in the field of biomaterials. With great uncertainty, evolutionary theory emphasizes the need for experimentation in a system. Technological (and scientific) diversity may therefore be considered as an indication of system performance as it presumably reflects the robustness of the system to the outcome of a selection process, and consequently, its growth potential.

As the technology is science-based, product development requires a great deal of time. Developing products for a medical device or pharmaceutical market requires clinical trials, and regulatory issues further delay market entrance, i.e., the *diffusion* process from the lab to the market is lengthy. Thus, an evaluation of the "closeness" to market exploitation was deemed to be appropriate. Rickne employed two different market-related measures of performance. First, she assessed whether or not the product had received regulatory acceptance by government authorities. Second, as the majority of the companies within this section of the biomedical industry need an agreement with a partner in order to have access to distribution channels, the number of partners was used as an indicator of closeness to market exploitation.

Finally, conventional indicators of the economic use of knowledge can be used, such as employment, sales, and growth figures. In addition, the financial assets the firms has managed to raise can be used as supplementary information of the ability to exploit knowledge commercially, indicating e.g. "staying power" as well as the interest in the firms from other companies or from the capital market.

To conclude, measuring the performance of a technological system is not straightforward but requires a careful consideration of the level of analysis applied and the degree of maturity of the technological system studied. Several indicators rather than only a single one are preferable, in particular when it comes to assessing the performance of an emerging technological system.

CONCLUDING REMARKS

This chapter has focused on the analytical and methodological issues which arise in the analysis of technological systems. We started by identifying the main theoretical dimensions of the concept: (i) technology clusters – the *cognitive* dimension, (ii) actor networks and infrastructures – the *organizational* dimension, and (iii) development blocs – the *economic* dimension. The importance of balancing and integrating these three components in the design and execution of empirical studies has been stressed.

These broad analytical concerns translate into a number of methodological issues. *Firstly*, what is the appropriate level of analysis for the purpose at hand? *Secondly*, how do we determine the population, i.e., delineate the system and identify the actors and/or components? What are the key relationships that need to be captured so that the important interaction takes place within the system rather than outside? *Thirdly*, how do we measure the performance of the system: what is to be measured, and how can performance be measured at the system level rather than at component level?

Given the special features of biotechnology, the analytical and methodological questions discussed in this chapter present a formidable challenge. The emerging technological system of biotechnology is very large, heterogeneous and dynamic. It is my hope that the ideas presented here will provide a useful framework for future research in this area.

Chapter 3

THE INDUSTRIAL POTENTIAL OF BIOTECHNOLOGY:
A Competence Bloc Analysis[1]

Gunnar Eliasson
The Royal Institute of Technology (RTH)

INTRODUCING THE MARGINALLY INFORMED ACTOR

This chapter addresses the problem of new industry formation in general – using competence bloc analysis (G. Eliasson and Å. Eliasson 1996) – and the industrial potential of biotechnology in particular. Is biotechnology one of the gateways to the new economy? This problem is large – to say the least – and incorporates a whole range of sub-themes, linking science to technology, technology to firm performance and then on to macroeconomic growth, addressing along the way the role of institutions, economic incentives and competition in economic growth, and finally policy. Can a meaningful role be identified for the policymaker in activating scientific results and technology industrially?

To address this complex problem in one comprehensive context we need good theory to organize our facts and thoughts coherently. Above all we need a dynamic theoretical framework without built-in answers to the policy problem, for instance making science and technology the drivers of economic growth by prior design.

Neo-classical and new growth theory as well as the linearized Schumpeter II (1942) model (by Freeman, 1974 and the Sussex School) suffer from this deficiency, however not the original Schumpeter I model of 1911, which I will call in for my analysis.

To achieve that I integrate three bodies of theory; first the notion of the knowledge based information economy (Eliasson 1990b) – very appropriate, indeed, for biotechnology, one of the few science-based industries (Eliasson 1998c) – to establish the vastness and extreme variety of the state space (or the business opportunity set) for competence bloc analysis. For practical purposes we define the opportunity set as the total knowledge of all possible, known and still unknown contributors of knowledge.[2] This means an enormous stock of knowledge accessible, except in fragments, to no single individual or group of individuals. It also means that the economy is operating far below its potential but that there is a great opportunity to learn, thereby automatically expanding state space. This definition is sufficient to lay the foundation of the main body of theory, or the Experimentally Organized Economy (EOE) in which economic growth is moved by experimental selection and learning in dynamically competitive markets through an (Table 1) entry, reorganization, rationalization and exit process.

Since firms are only marginally informed about the business opportunity set, they make frequent business mistakes during their search into it, but now and then come up with winning solutions. This search by millions of firms is a learning experience for all, which keeps shifting the state space or the business opportunity set outwards; possibly at a rate that is faster than the actors can learn, making them increasingly ignorant about all they can learn (cf Information Paradox I on p. 46f in Eliasson 1990b). The search by actors who are grossly ignorant about most has two dimensions; (1); even if your business idea may make you confident that you are right, you know you may be completely in error; and (2) even though you are grossly ignorant as an individual, all ignorant actors together may possess sufficient knowledge to create and develop a new industry. The important thing is to support an organization of the economy such that the total of the competencies of all actors is efficiently allocated to that effect. We know that the payoff to such a successful outcome is very large for all participating actors. Hence, there should be a private demand for the endogenous development of such an organization of competent actors that this positive collective effect is achieved. The economy faces an interesting dynamic positive sum game. Has the policymaker a role in catalyzing that game?[3]

We can make an empirical statement already here. Even though the characteristics of the EOE are everywhere present, they are particularly appropriate for the new generic technologies like computer and communications and biotechnology. Generic technologies cut deeply into the production structure and carry the potential of fundamental reorganization, like the machine tools in the first industrial revolution. When commercializing biotechnology the individual company enters an enormous business opportunity set. For companies in the EOE, however, this is the normal experimental situation. They have to try out their business ideas (test their business hypotheses) in the market for progress to occur. For the economy at large the business mistakes that occur along the way should be regarded as a normal cost for learning and economic development.

For the central policy maker, on the other hand, the situation is different. He commands too large resources, and exercises too large a leverage on the entire economy to be allowed to carry out policy experiments, risking failure. The charter of the public policymaker is to do it right ("the judge"). He or she, therefore, embodies an inclination not to accept a mistake and to be unwilling to correct it that often makes him or her incompetent for market activity (Eliasson 1996a, p. 57). I will return to this.

A critical policy problem, however, is to make sure that the economy is well organized for business experiments in the market such that the incidence of two types of errors is minimized, i.e. (Table 2) such that losing projects are not allowed to go on for too long and that winning projects are not lost.

How can this be achieved when everybody is grossly ignorant and possibly mis-informed about what is relevant, and nobody can define the knowledge that is needed (it may be tacit). For this you need institutions supporting efficient organization, functioning economic incentives and competition. Competence bloc theory brings it all together within the dynamic context of the EOE.

COMPETENCE BLOC THEORY

Competence bloc theory begins from the demand side, the customers setting the criteria for economic selection and important competence contributors. In the long run there will be no better products developed and marketed than there are customers willing to pay.

The competence bloc defines the minimum set of competencies needed (through its actors as carriers of competence) to initiate and develop an industry defined as complementary products and competing substitutes. With the customer being the ultimate arbiter of the selection of projects also the corresponding industry should be defined and delimited from the demand side.[4] Development occurs through selection and learning in markets. Critical competencies are often tacit and impossible to define. Hence, they are represented by their carriers rather than functionally defined, leaving open great spaces of individual ignorance. Since actors are differently ignorant the total mass of knowledge or competence in the economy may still be very large. One task of the competence bloc is to activate and allocate that total competence mass efficiently. The technology or competence input in the competence bloc is what you also need to initiate and develop an industry. But what you need is not known or predictable before the creation of a new project and its selection in the competence bloc has taken place. For that selection to be efficient far more technological variety has to be developed (supplied) by innovators than is eventually selected as commercially winning projects.

A well-organized competence bloc coordinates carriers of tacit competence such that each project is subjected to a maximum exposure to a varied and competent economic evaluation, such that the two errors in Table 2 are minimized. More particularly, in a well organized competence bloc potential winners can keep experimenting, being confident[5] that they will eventually come out as real winners. Continued search is subjected to increasing returns for potential winners. A positive sum game is created (Eliasson and Taymaz 1999). Competitive selection moves the entire bloc. Growth occurs.

Obviously, then, the existence, completeness and the organization of institutions supporting an efficient competence bloc will be a factor to consider in industrial policy analysis. Furthermore, for a science founded industry like biotechnology, university entrepreneurship will be a critical complementary policy issue.

The Theory

The role of competence bloc theory (G. Eliasson and Å. Eliasson 1996, Eliasson 1997a, 1998a) is to explain the competitive creation and selection processes that generate growth in the experimentally organized economy (Eliasson 1991, 1996). Its basic rationale is that it is more important economically to do the right thing than to do it efficiently. Hence, customer satisfaction and (to be shown) customer competence contribution are in focus. While the traditional Walras-Arrow-Debreu model "embodies" only one (optimal) equilibrium solution without choice, the EOE offers an incomprehensible variety of choices and ways of organizing this selection. Competence bloc theory is an analytical device to explain this organization and the development of an industry driven by the complex interaction of competent actors, the competencies of - whom to perform particular tasks (functions) cannot be defined (specified) as to content, only be characterized as to results (output). The minimum set of competent actors of the competence bloc is exhibited in Table 3.

EOE and competence bloc theory together define a dynamic process that explains how the technologies needed to build a new industry are:
- created (innovation)
- identified (recognition)
- selected (competition)
- commercialized and diffused (market support) and
- competently introduced in production (receiver competence)

such that the right (product) technology choices are made and two types of errors (Table 2) are minimized, i.e. (1) to keep losers for too long and (2) to reject winners. In an efficiently organized and well staffed (with competence) competence bloc winners face increasing returns to continued search. In the selection process two types of phenomena appear; (a) synergies and spillovers arise (Eliasson 1997a, 1998a, b, c). This is all a matter of applying the right diversity of competencies in the selection process, (b) business mistakes appear as the necessary consequences of a learning process and figure as a standard cost for economic development (Eliasson 1992).

Table 1. The four mechanisms of growth in the experimentally organized economy

1	Entry
2	Reorganization
3	Rationalization
4	Exit (shut down)

Source: "Företagens, institutionernas och marknadernas roll i Sverige", Appendix 6 in A. Lindbeck (ed.), Nya villkor för ekonomi och politik (SOU 1993:16) and G. Eliasson (1996a, p. 45).

Table 2. The dominant selection problem

Error Type I: Losers kept too long
Error Type II: Winners rejected

Source: G. Eliasson and Å. Eliasson, 1996. The Biotechnological Competence Bloc, Revue d'Economie Industrielle, 78-4^{0}, Trimestre.

Table 3. Actors in the competence bloc

1. Competent and active customers
2. Innovators who integrate technologies in new ways
3. Entrepreneurs who identify profitable innovations
4. Competent venture capitalists who recognize and finance the entrepreneurs
5. Exit markets that facilitate ownership change
6. Industrialists who take successful innovations to industrial scale production

Source: G. Eliasson and Å. Eliasson, 1996. The Biotechnological Competence Bloc, Revue d'Economie Industrielle, 78-4^{0}, Trimestre.

The selection process in the competence bloc (through Table 3) is organized as follows:

First, the products chosen or created in the process never get better than what customers are capable of appreciating and paying for. The long-term direction of technical change, therefore, is always set by the customers. This is so even though the innovator, entrepreneur or industrialist takes the initiative. But quite often the customer takes the initiative. Technological development, therefore, requires a sophisticated customer base (G. Eliasson and Å. Eliasson 1996). In one sense, the customer analysis of competence bloc theory opens up the Keynesian macro demand schedule. But as you peek inside that "black demand box" you will find that the customer dynamics of the competence bloc has little to do with Keynesian demand. The actors of the competence bloc contribute (commercial) competence in the technological choice process. This argument also serves as a rationale for competent purchasing and acquisitions, including public competent purchasing in areas where goods and services are supplied by public authorities. In short, to enhance product value, the selection and production processes have to be focused on the customer.

Second, basic technology is internationally available, but the capacity to receive it and make a business of it requires local competence. Part of this receiver competence (Eliasson 1990a, 1996, pp. 14 ff, pp. 212 f) is the ability to create new winning combinations of old and new technologies (innovation). As we know (see e.g. Larsson, Lembre, and Mehldal 1998) a rich and varied supply of subcontractor (technology) services is part and parcel of an efficient innovation process.

Third, the task of the entrepreneur is to identify commercial winners among the suppliers of innovations and to get his/her choice of technology (technology choice) on a commercial footing.

The Financier Does Not Understand The Entrepreneur

The entrepreneur, however, rarely has resources of his own to move the project forward. He or she, therefore, (fourth) needs funding from a competent venture capitalist, i.e. a provider of risk capital, capable of understanding innovators of radically new technology and be able to identify business needs and provide context. The money is the least important thing. What matters (Eliasson 1997b, G. Eliasson and Å. Eliasson 1996) is the competence to understand and identify winners and, hence, provide reasonably priced equity funding.[6] The supply of such competent venture capital is extremely scarce. It is the critical part of the overall selection process and, if lacking in performance, is liable to result in the "loss of winners". In fact, completeness is a necessary condition for a functioning competence bloc. Making the competence bloc complete must, therefore, be the prime task of industrial policy (Eliasson 1998c). None of the "pillars" (the agents) of the competence bloc can be missing, or the whole incentive structure will fail to develop (G. Eliasson and Å. Eliasson 1996). The venture capitalist, and his escape (exit) market (fifth) are the most important incentive supporting actors. With no understanding venture capitalists the price of new capital will be prohibitively high or not available, and winners will be filtered away. With badly functioning exit markets the incentives for venture capitalists will be small and, hence, also for the entrepreneurs and the innovators. Completeness of the competence bloc is, therefore, a necessary requirement for the viable incentive structure that guarantees increasing returns to continued search for winners, i.e. for new industry formation. The venture capital market in Sweden is generally lacking in the industrial competence needed to fund radically new industry (Eliasson 1997b), and even though the exit market situation has been improving, compared to the US, Sweden is still an underdeveloped economy on both counts. It is, hence, very risky to be an innovator and entrepreneur in Sweden, since when the two have exhausted their own resources, there will be no one to turn to except unappreciating bankers, big company executives or public sources, all more or less incompetent in dealing with radically new industrial ventures. The risk is high that winners will get lost.

Finally and sixth, when the selection process has run its course and a winner has been selected a new type of industrial competence is needed to take the innovation to industrial scale production and distribution. We cannot tell in advance what the formal role of the industrialists is (CEO, chairman of the Board, an active owner etc.). He or she figures in the competence bloc on account if his or her capacity to contribute functional competence.

It is true that, for instance, Sweden features an extreme concentration of large-scale business leadership competence (Eliasson 1996, 1998d), but this competence has been acquired in traditional, mature industries that innovate slowly. The management of innovation in health care and biotechnology industries is radically different from that in mature industries like engineering. The general experience is that leadership competence acquired in traditional industries is of limited use in the radically new industry that has to be created for a mature industrial economy to transform itself successfully towards its potential. (See Eliasson 1997b, 1998c, Hellman 1993, Henrekson and Johansson, 2000).

A viable competence bloc now has to exhibit two dominant properties:

(1) Increasing returns to continued search (R&D); if a winner enters the competition race, continued search (innovation experiments) will result in commercial success.

(2) Sustained incentives; the competence bloc has to be complete to support incentives for such sustained search (completeness).

As we have already noted, completeness is a tough requirement. If one link in the experimental selection process of the competence bloc is missing or faulty (for instance because a competent venture capital industry does not exist) incentives will not be sufficient to stimulate the necessary innovative activity. These systems properties will exhibit themselves when the competence bloc is complete and when sufficient critical mass has been achieved. Then the competence bloc will function as an attractor such that new entry takes place in such a way

- that the competence bloc benefits from the new entrants, but also (because of competition) such
- that only the new entrants that contribute to the competence bloc enter and/or survive.

Then the competence bloc will begin to develop endogenously through its internal momentum (critical mass). We have a dynamic positive sum game.

As a consequence of these synergies and the diversity (pluralism) of approaches and agent representation the allocation and use of the existing competence mass will be optimized and spillovers will characterize the developing competence bloc. These spillovers will diffuse along many ways and both further reinforce the internal development forces of the bloc and contribute serendipitously to other related and unrelated industries (Eliasson 1997b).

Technological Diffusion

The diffusion of new technology occurs along six distinct channels; (1) when people with competence move over the labor market, (2) through the establishment of new firms when people with competence leave established firms, (3) through mutual learning among subcontractors and the systems coordinator, (4) when a firm strategically acquires other firms to integrate their particular knowledge with its own competence base, (5) when competitors imitate the products of successful and leading firms (the "Japanese approach") and (6) through organic growth of, and learning in incumbent firms. (Eliasson, 1995)

One important insight emerges from a close study of the above; efficient diffusion of new technology requires effective market support, notably in the labor market (item 1 above) but also in the venture capital market and the market for mergers & acquisitions (M&A, see further G. Eliasson and Å. Eliasson 2000). Efficient diffusion is also a necessary condition for spillovers and competence bloc development, but it is not sufficient. For new technology to be introduced in production receiver competence (Eliasson 1990a) is needed. Entrepreneurial and venture capital competence are part of this, but the general and rapid introduction of new technology also requires a varied and competent labor force at all levels (workers, engineers, managers and executive people). Thus, when we integrate institutions, incentives and competition into one coherent theory we arrive at the full model of the experimentally organized economy, featuring growth through competitive selection as in Table 1.[7]

BIOTECHNOLOGY AS A NEW INDUSTRY[8]

Biotechnology is an old industry with beer and wine making as early folk practices and sub-industries. In its modern, science based form it is, however, it is very young, beginning with the (Watson and Crick 1953) discovery of the structure of the genetic information base as a Double Helix of DNA. This suggested to them how to copy the genetic information contained in the DNA chemically, an hypothesis that rapidly led to several path breaking scientific discoveries around 1975 that in turn have spun off several derivative methods for commercialization (see Table 4). It should be observed that biotechnology uses biological science, but exists mainly in the intersection of biology and chemistry (biochemistry).

Table 4. Basic biotechnology

1.	rDNA (including cloning, sequencing and polymerate chain reaction) technology[a] (Boyer and Cohen et al., 1973)
2.	Protein engineering[b], including X-ray chrystallography
3.	Immunology, for instance the use of antibodies and diagnostics (Köhler and Milstein 1975, Nobel prize 1984)
4.	Fermentation, cell culture and volume production of biological substances, including embryo rescue and in vitro fertilization
5.	Filtering and selection technology, using i.a. gels and columns (Pharmacia, Uppsala)
6.	Combinatorial chemistry and other related techniques.

[a] rDNA stands for recombinant Dioxiribo Nucleic Acid.

[b] rDNA technology and protein engineering have such a large common intersection that it is often difficult to keep the two technology areas apart. For examples, see Eliasson, Å. (2000).

Source: Eliasson, Åsa, 2000. A Competence Bloc Analysis of the Economic Potential of Biotechnology in Agriculture and Food Production, Working Paper IBMP-CNRS, Strasbourg and KTH, Stockholm.

Biotechnology is special also in other ways. As an industry it is commonly defined in terms of its molecular technology components which is a first source of problem, since technologies change continually all the time. Second, an industry and a competence bloc need an end product, market value added definition. This is almost impossible for the output of an activity (biotechnology) that is deeply integrated with the final user activity, for instance diagnostics in health care. There is also the important problem of distinguishing between the output of the industry as a new product/service and as a process improvement. But this is a common problem with all industry definitions. More problematic is the intermediate character of any definition of biotechnology industry as a supplier of input products or technological services to other industries like pharmaceuticals/health care (see Table 4), agricultural and food production or pulp and paper industry. There is no well defined market for the products of biotechnology industry.

To clarify, let us look at biotechnology as having three distinctly different dimensions;

(1) as a generic technology

(2) as an industry producing biotechnology goods and services

(3) as a broad based user industry composed of all the demand defined competence blocs which use (1) and (2) as inputs, health care (Figure 1) being one such competence bloc.

As a rule the customer of these buying industries are either very professional or professional but somewhat special. Agriculture and food industries face a particular problem. First of all, farmers will be buying high-tech supplies, the nature of which they do not understand. The products they in turn put to market are sold to even more ignorant customers and/or by way of their substitute (representative) customers in the form of regulators and politicians who are presumed to know better. Transgenic food is a good illustration (more on this below).

Pulp industry is even more interesting. Pulp making technology is dominantly chemical. The introducers of biotechnology in pulp making should, therefore, be very professional customers, but their competence is in the wrong field. The traditional technology of pulp production is to remove undesired components from the cellulose fibers through chemicals and energy intensive methods ("cooking"). The new biotechnology is also facing conservative resistance from the chemical engineers in this industry (see Laestadius 2000).

The point of the argument is that biotechnology firms face no well defined traditional market; and if they do, they operate far away from their ultimate customers. Health care illustrates (Figure 1). Here biotechnology enters as a technology supplier to pharmaceutical industry and directly as advanced treatment methods into the hospital care industry. It is difficult to identify well defined products for this industry.

DEFINING BIOTECHNOLOGY BY ITS INGREDIENT TECHNOLOGIES

Watson and Crick (1953) laid the scientific foundation of a sequence of laboratory techniques around 1975 (see Table 4) developed at the time for university laboratory use only (Å. Eliasson, 2000) that were later commercialized as industrial applications. There are at least six basic stepping-stones into industrial technology. All techniques in Table 4 except 3 (Immunology), 4 (Fermentation etc.) and 6 (Combinatorial chemistry) derive from the old biochemistry field.

rDNA technology including cloning technology is currently the most fashionable biotechnology area. It includes as a further development transgenic plants or genetically modified organisms (GMOs).

Immunology, for example the use of antibodies, features many medical and pharmaceutical applications, notably in the diagnostics area. The production of vaccines in plants (see Eliasson, Å. 2000) is another example.

Protein engineering includes a number of useful molecularly based techniques that include X-ray chrystallography and molecular computer modeling. The products and production of Swedish KaroBio are based on these techniques (see below). Protein engineering cannot, however, be clearly distinguished from rDNA technology. Most protein engineering methods use rDNA technology in one way or another.

Furthermore, fermentation, cell culture and volume production of biological substances are a rapidly expanding field of bioindustrial applications. We have the production of enzymes for detergents, biodegradable chemicals and hormones. The large-scale production of all kinds of biological substances uses these techniques. We also have in vitro culture of plants, in vitro fertilization and embryo rescue and regeneration of plants (that integrate 1 and 4 in Table 4).

Finally, we have the important supporting filtering and selection technologies of early biochemistry without which many of the other biotechnologies would not have been practically feasible. This development began (in Uppsala) already in the 1950s, and may be said to be the foundation of Pharmacia[9] as an industrial high technology firm. Pharmacia in Uppsala played a decisive role in supplying such filtering and selection technology to the development of the entire biotechnology area during the late 1970s and the 1980s. In fact, the early venture capitalists in the area confused Pharmacia with a "true" biotechnology company and believed that the portal firms in the biotechnology revolution were Pharmacia (Uppsala) and Novo, Copenhagen (Eliasson 1997b). The latter had developed a method (under 4 in Table 4) to produce raw penicillin.

Combinatorial chemistry and other related subfields are direct applications of organic chemistry that are frequently used in combination with the other techniques in industrial applications. For instance, the Uppsala firm Personal Chemistry (previously Labwell) has developed a technique to fasten chemical reactions, using microwave technology. Earlier it could take days, sometimes forever, for a pharmaceutical firm to identify a substance (a molecule). With the new method we talk about minutes, and also the possibility to simply make thousands of substances that would otherwise not be feasible to synthesize due to their complex nature.

In general, industrial applications of biotechnology mean integrating different technologies. Thus, for instance, the use of plants as plants to produce color pigments or oil (for instance algae to produce petroleum, see Eliasson, Å., 2000) or pulp (Laestadius, 2000) means integrating rDNA technology and fermentation and cell culture (the technology). Often the same result can be reached by many different methods. The forest industry might want a particular kind of wood to be used in pulp production. This is normally done through traditional cross breeding and selection, possibly speeded up by, for instance, chemicals or ultraviolet (UV) radiation ("artificial mutagenesis"). Forest or pulp industry, however, can also use rDNA technology (item 1) to produce genetically modified trees that grow better in cold climates and/or are modified to better suit the chemically based pulp making process.[10] However, one has difficulties seeing a principal difference between changing existing genes through subjecting them to strong chemicals and replacing one gene with another through genetic engineering. In fact, rDNA technology puts biological knowledge to more systematic and controllable use. Functional food also builds on the integration of rDNA and cell culture technologies and the production of vaccines in (for instance) fruit involves the integration of rDNA technology (item 1), immunology (item 2) with cell culture techniques (item 4).

We can describe the production of the Swedish biotechnology company KaroBio (see Eliasson and Eliasson 1997) in a similar fashion. In pre-clinical testing substance screening KaroBio integrates rDNA technology (item 1) with computer-aided drug design and crystal structure analysis (item 3), tissue profiling (item 4) and combinatorial chemistry (item 6) with immunology (item 2).

All these combinations point to the difficult problem of defining an industry in meaningful economic terms. Statisticians have attempted to identify stable industrial characteristics that relate one to one to the output value added of the particular industry being characterized. In the past this meant inputs (pulp, chemicals) or production processes or activities (engineering) and more recently technology (IT, biotechnology). Engineering industry is both a technology input definition, based on the machine tool technology that once formed the basis for the industrial revolution and a production process definition using these tools. But this relates very weakly to final use and the customers. And the statistical engineering industry definition also provides us with a hopeless spectrum of the most widely different products, including everything from woodsaws to automobiles, airplanes and computers. The choice of basic characteristics for the industry definition is no trivial matter, and for two reasons; (1) Once chosen you tend to get stuck with the definition and (2) once established its basic deficiencies as a proxy for output (the validity problem) tend to be commonly neglected, the measure being used ignorantly for analysis, provocatively in politics and dangerously as support for policy making (see for instance on the erroneously conceived "de-industrialization debate" in Eliasson 1990b, pp. 44f). Biotechnology provides an excellent reason for trying to shift the industry definition forward to its demand categories. None of the technology characteristics (using its molecular technology components as a base) are sufficiently stable to serve as valid proxies for the industry. The Watson and Crick "discovery" is too general and still does not cover all the old biotechnology based industries like beer and wine. Going from Watson and Crick (1953) to Table 4 makes the categorization unstable. The Cohen and Boyer et al. (1973) and the Köhler and Milstein (1975) achievements resulted in laboratory methods designed for the scientific ("university") laboratory. These researchers had no industrial applications in mind. First of all, we are now talking about laboratory methods based on the propositions derived from Watson and Crick's (1953) genetic copying theory. There may be many more different and better such scientific laboratory techniques. Second, the industrial techniques that started the biotechnology revolution were derivative methods from these scientific methods, that were different for each application. Thus, for instance, cell culture (item 4 in Table 4) for bacteria and plants may be based on the same scientific methods but their industrial applications are based on widely different technologies. Third, and finally, industrial technologies tend to crossbreed, as we have shown, through the categories of Table 4, and there is no guarantee neither that the categories will be the best ones ten years from now, nor that new items won't be needed in the table, that come from entirely different fields. While the demand (customer) industry definition is theoretically correct it still holds difficult statistical problems. On the other hand, while the input or process definition may be easier to deal with statistically it does not lend itself to answering interesting problems and it may give misleading answers. It is no coincidence that when business firms divisionalize to understand their interior production processes and sources of profits better they use market criteria, not input or technological criteria. So the most important piece of advice is to move the definition closer to the customer as the business firms do. They look at their market share and the total market is defined by complementary products and competing substitutes.

It is true that even though theoretically clear, this is also a difficult definition empirically to use. It is, however, always to be preferred to use a theoretically clear and desirable but difficult definition to guide measurements than to have a theoretically unclear and difficult to measure definition. There is no way of avoiding, in deciding on statistical nomenclatures, a certain amount of arbitrariness. The clear theoretical definition, however, helps to evaluate the errors we make. Hence, a biotechnology industry has no stable technology (input) definition, but this discussion, nevertheless, gives advice on how to define an industry.

This discussion of a technology-based industrial classification raises and illustrates three problems; the generic instability of any technology classification, the difficulty of defining an industry by its ingredient technologies and the fast deterioration of meaning of any technology concept above a very precise micro level specification. For instance, the possibly very important industrial applications of producing vaccines in plants combine three different sub-technologies, and the way they are integrated depends on economic circumstances (relative prices, productivity factors etc.). Similarly, protein engineering (item 3) increasingly employs computerized molecular design techniques etc.

Biotechnology, hence, rapidly fragments into a spectrum of sub-technologies that are integrated differently for each industrial application, that increasingly integrate into other technologies external to biotechnology, like chemistry, computer science etc.

Biotechnology, furthermore, offers many different ways to achieve the same ends; a new product or new properties with a vegetable.

This forces us to define the meaningful economic categories at a different level, far away from the technology and closer to the market. This is also one idea of competence bloc theory.

DOUBLE FOCUS DYNAMICS

The many dimensions of technology discussed in the previous sections introduce interesting dynamics into competence bloc analysis. A user-industry is composed of a technology supply (technology systems) side (as Bo Carlsson has discussed in the preceeding chapter) and a demand (competence bloc) side; for instance the health care industry in Figure 1.

Since the technologies being selected and traded are basically tacit we have to insert (between the two) a bloc of acting intermediators. There the entrepreneurs of the competence bloc meet and evaluate the innovators of the technology system. Together the integration of these three and extremely complex dimensions generates the multidimensional dynamics that ultimately -moves economic growth, and the outcome of which is very hard to predict.

Biotechnology supplies a great variety of technologies, goods and services for the health care industry. These technological products are sorted, combined internally, and with external technologies eventually to enter, for instance, health care as indicated in Figure 1. At some fairly low level of aggregation we can talk about a technology system (Carlsson 1995, pp. 7f, 23, 49f). As the particular and well defined technologies are filtered into the competence bloc and combined with other technologies the composite technologies become increasingly "economic" in character, and unstable as to their composition. Biotechnology, hence, enters the competence bloc for filtering in the form of different technologies or different technological systems under the innovation item 2 in Table 3. The competence bloc can accommodate and filter a large number of specific biotechnologies. They enter in the form of a spectrum of proposed innovations. Only some of them are, however, commercially viable and the principal idea of the entrepreneurial function 3 in the competence bloc is to select a subset of economically viable (read profitable) innovations to be moved further up the competence bloc. This means that a viable technology system should be capable of generating great and innovative technological variety for economic sorting and selecting in the confrontation with the ultimate users (demand, customers) in the competence bloc. The intersection between the innovative and the entrepreneurial items, hence, defines a dynamic and important market where innovations are supplied in the form of differently integrated technologies (or technological systems) for industrial evaluation by (ultimately) the industrialists and customers of the competence bloc, eventually to manifest themselves in the form of value creation, or growth in the corresponding industry.

OUTCONTRACTING

Large firms typically do not engage in radical innovative activity, even though they may be carrying on very efficient operations that are constantly improved upon. Sometimes, radical innovation is called for also in the large firms. Their technology base may be killed by a competitor. This has been common in computer industry, and the large firms in that industry have had great difficulties in coping with radical technological change. Sometimes also the big firm reorganizes radically to gain an edge on its competitors, for instance, by building a new type of international marketing organization. The large Swedish firms were very successful here during the 1980s (Eliasson 1990b). General Electric, a giant and very old firm, has been admired during the 1990s for its organizational dynamics. Small firms, on the other hand, are typically innovative in the traditional sense. They would not survive otherwise. The small firm with the great product, on the other hand, doesn't have the resources of the big pharmaceutical company to carry out extremely risky and resource demanding clinical tests and to take the product to the global market through an enormous sales network. This dichotomy is particularly dangerous in pharmaceutical industry, the big firms of which need a steady creative supply of new innovative and commercially viable substances to survive in the long run. This dichotomy also mirrors the distinction between doing (selecting) the right thing and of doing what you do right. While the small firm, being deeply submerged in its particular technological know-how may have difficulties sensing the customer appeal of its particular, often very intermediary innovation, the big process-oriented firm, even though closer to the market and the final customers may have difficulties, as well, understanding the appeal of the new innovative product with its traditional customer base. For the necessary and efficient sorting and the varied competence exposure you need a large number of innovators and the varied competence supply (for evaluation) of the actors of the competence bloc. Apparently there is a need for decentralized cooperation, and sophisticated outsourcing is one solution that is endogenously developing in the market. Swedish KaroBio (Eliasson and Eliasson 1997) has identified one such niche between the university and the large pharmaceutical company. It brings commercial experience and competence, and also venture financing to the university lab by helping to identify commercially promising projects. KaroBio, thereby, supports university entrepreneurship. It integrates rDNA technology with computer-aided drug design, crystal structure analyses, tissue profiling and combinatorial chemistry with immunology for pre clinical (testing) screening of proposed drugs, and then markets a narrowed down selection of evaluated substances to the large pharmaceutical company for further clinical testing. From the point of view of the large pharmaceutical company this means that the early innovative research stages have been outsourced to the university and to KaroBio. This is very good since the large companies are not very good in handling the early innovative stages of radically new product development. Outsourcing of advanced product development work in turn requires specialist competence. It is, however, a general organizational technology with very large potential productivity effects. The exact format and judicial form of this outcontracting depends on financing arrangements, the sharing of risks and (above all) on the existence of sophisticated markets for strategic acquisitions (Eliasson and Eliasson 2000).

STRATEGIC OBJECTIVES AND THE FINANCING AND RISK SHARING ARRANGEMENTS OF NEW START UPS

The objectives of new start ups are not independent of their sourcing of finance and the agreements on risksharing. There are three principally different ways of funding innovative commercialization of biotechnology (see Eliasson, Å., 2000).

I High risk ambition:
Build the company on your own to industry level on external venture financing.

II Medium risk ambition:
Aim for pilot product being strategically acquired by large company.

III Low risk ambition.

Do Contract Work for Large Company?

The first high-risk ambition requires the support of a complete competence bloc. The second, medium risk strategy requires the presence of fully developed markets for strategic acquisitions, notably for bidding up prices of acquisition objects such that incentives for innovators are sufficiently high. In the highly competitive Internet market such markets have developed automatically. Telecom systems developers which did not have the needed particular broadband and server technologies in 1999 paid anything for small companies possessing them to acquire the needed technologies. The situation is not as good in biotechnology where acquisitions sometimes barely mean cost recovery for the innovator firm. This is partly because the component ("modular") technologies have not been fully standardized and developed, but also because a sufficiently large number of competing and competent buyers are missing.

Each of the three strategies means a different sharing of risks between the innovating company, the financing intermediator (strategy I) and the user of the technology. It is interesting to note that the strategic acquisitions market and the market for contract work are two different forms of very sophisticated outcontracting of research by the large firms.[11] Outcontracting contract work means that the buying firm knows what it wants (specifies the work) even though the contractor may have been marketing its ideas. It usually supplies unique competencies and licensed technology. The low risk, contract work strategy means that the firm does not feel itself sufficiently competent at the entrepreneurial level to take on large risks, but rather prefers to focus on its scientific, technological and innovative competencies (item 2 in Table 3). In the market for strategic acquisitions the small innovative firm takes the initiative expecting that there will be a demand for what it comes up with. It has to possess both innovative and entrepreneurial competence in terms of Table 3. Both ways research is done that the big buying firm could have carried out internally in principle, but much less innovatively and less efficiently, and often without a winning outcome.

If the little firm opts for the high-risk strategy I it must feel competent also at the industry and final customer levels in Table 3. While the medium risk strategy II (above) means that the innovative firm takes the risk that there are no competent strategic buyers the high-risk strategy I means that the firm risks facing an underdeveloped and incompetent venture financing industry (items 4 and 5 in Table 3),

The nature of risk taking, hence, is very different at the three levels and has to be partly subjectively determined. It is a composite of risks associated with the lack of competencies (understanding) throughout the competence bloc. It is asymmetric. What looks highly adventurous to one actor is under full subjective control by the other.

The low-risk contract work strategy means no risk taking except that there may not be any large company that understands the particular innovative competence of the contractor firm to solve its technical problems.

The medium risk player faces a possibility that there are no competent (by its definition) strategic buyers. It has to market itself widely to find the one which pays very well. Otherwise it will be picked up cheaply, and be wasted, or somebody else will take the money. This observation is critical for the development of a sophisticated intermediary and difficult to define industry like biotechnology. We are talking about the existence (and where) of sophisticated customers and there have to be several "within reach". That is the gamble of the little firm.

Again the high-risk strategy I takes on the additional risk of also moving the entire project through its venture financing (competence) stages to full scale industrialization, or almost so. If the innovator/entrepreneur feels that it possesses the whole range of competencies (in the competence bloc) needed for this its risk is subjectively low, except it must take the risk of an insufficient supply of understanding venture capital.

To deal with this situation analytically we have to face up to some tricky problems indeed: We are not studying calculable risks in the sense of standard financing literature. For the whole of a risky project we face a composite of calculable and subjective risks that together have all the characteristics of uncertainty as distinct from calculable risks. While the risks associated with a deficient venture capital competence can probably be objectively calculated to determine an optimal financing[12] and marketing strategy, the competence bloc is rather an organizational solution to decentralize or outsource, not only the evaluation of projects but also the risks associated with each component evaluation. The composite total of risks is not a tradable item ("risk") in the sense used in financial economics. We are talking of the decentralization of the absorption of uncertainty, and any tradability of that uncertainty will have to rest on an assessment of the competence of its embodiment (its "carrier") in the competence bloc. This is not a negligible circumstance to consider in the study of project selection and in business management. Normally the project itself is not evaluated but the competence of the person or the group of individuals who have initiated the project and done the screening. This is what I call "management by approximate skills" (Eliasson 1998d) that is unavoidable when it comes to evaluation by tacit competencies. In the venture financing context the venture capitalist (referred to above) brings together a group of people to do the evaluation and often a complex and/or radically new project passes several groups for such evaluations. It is thus easy to understand that the existence and availability (access to) such groups of competent people are critical for success in capturing winners. In the exit market this (Eliasson 1997b) type of "proxying" is even more important. The venture financing now is to be unloaded to ignorant large investment funds or the public. For such IPOs to be accepted they have to be emitted by an investment firm with a brand name that reflects trustworthiness. The policy and competence of such investment firms also are to build such a reputation to be embodied in its name.

As a rule the three forms of financing are mixed strategically. The entrepreneur in biotechnology might want to go all the way on his/her own, but sells out when offered a good price. Similarly, an entrepreneur à la strategies I or II may be offered a generous contract when a large pharmaceutical firm realizes it needs what the innovative firm is doing and wants to lock it up before a competitor does.

HOW MATURE IS BIOTECHNOLOGY INDUSTRY?

Biotechnology introduces both (1) radically new methods in the production of traditional products and (2) radically new products. The same is true of computing and communications (C&C) industry where previously mechanical technologies have been replaced by "digital" technologies.[13] In the latter C&C industry each new technology generation has tended to kill the previous generation. The same radical change in processes and products has not occurred in engineering industry which still features significant elements of the production technology that once ushered in the industrial revolution; machine tools. There is, however, a difference between C&C industry and engineering industry. C&C technology is generic and directly influences (integrates with) almost all production. Engineering, on the other side, is an industry that draws on a large number of technologies integrated in products and processes through a dominant organizational competence. IT or C&C technology is one such component technology. Here, again, engineering industry distinguishes itself from large-scale process industries like steel, pulp or paper that are not moved by organization and a dominant technology. Again it is possible to envision a radical destruction of chemically based pulp manufacturing if a radically new and superior biotechnology is developed (Laestadius 2000). The question raised in this section is whether those particular characteristics of each industrial category are typical of the particular product and process technologies or if the various industries are constantly changing in a common direction, for instance, C&C and biotechnology gradually taking on the organizational characteristics of engineering technology. Thus, the increasing importance of specialized software development and diversified marketing in C&C industry is making organization changes important there. In biotechnology volume production of new penicillin was an early development, but the advanced protein engineering of extremely varied output qualities that is characteristic of pharmaceutical industry and similar users of the new biotechnology are taking it in a direction closer to that of engineering industry.

The production organization patterns that the new biotechnology based industries are adopting are the results of very competent and innovative combinations of science, experience and industrial management. The complexity is illustrated by the fact that large-scale management experience from engineering industry apparently does not lend itself to easy "cloning" in biotechnology or pharmaceutical industry. One indirect illustration of this is that Sweden seems to have exhibited industrial success in areas where biotechnology has been integrated with mechanical and/or electronic technology like in Gambro (dialysis), (Siemens) Elema (pace makers), Electa (precision radiation brain surgery) etc. The integration of mechanical devices with electronics and with new materials in new products has been a successful characteristic of post-war Swedish industry.

But again, biotechnology and its user (demand) industries are largely different, drawing among other things heavily on an academic infrastructure. The importance of a network of subcontractors was repeatedly emphasized in our interviews of newly started biotechnology companies in the Uppsala area. More recently the importance of industrial organizing competence made available for the same companies because of the break up of Pharmacia research and production in the area after its merger with UpJohn has been emphasized. These are signs of a mature or maturing industry, i.e. specialized networks of suppliers (subcontractors) and standardized management experience that can be moved from firm to firm. It was also emphasized during several interviews (see Eliasson and Eliasson 2000) in the Uppsala area that the availability of specialist subcontractors in the neighborhood made it possible for newly established firms, based on a new product technology to move (grow) fast since they did not have to invest in, and fragment their management attention on everything. The next step in the same direction is the establishment of a diverse network of specialized new technology suppliers and a parallel diverse market for strategic acquisitions (many buyers), all typical of a full-bodied competence bloc like the C&C competence bloc in Silicon Valley.

POLICY IMPLICATIONS

The 1960s bred a large audience of optimistic industrial policy makers using, increasingly during 1970s and part of the 1980s, the Japanese "success" as their guiding example. The linear Schumpeter (1942) II technology growth model, notably presented by the Sussex School of technology economists (see, for instance, Freeman 1974), complemented Keynesian demand policy in shifting emphasis on industrial policy from private to central Government initiatives. Technology was what moved economic growth and all that was needed was to provide resources for technology. At the time that meant R&D, notably university R&D. There was no discussion about the competence of investing in the right R&D projects, and the risk of failed projects or lost winners. The underlying economic rationale was the mainstream general equilibrium model, finally polished up for mandatory academic teaching during the 1970s.[14] Japan, and MITI's great focus on biotechnology[15] was the great scare for European and the US participants at a very interesting technology conference in 1981 at the Kieler Institute under the chairmanship of professor Herbert Giersch (see for instance Rehm 1982). The fact that Japan developed no subsequent significant innovative presence in biotechnology carries significance for my policy discussion. It has to do with the enormous complexity of industrial technology in general, and biotechnology and C&C technology in particular that invalidates completely the central overview argument for industrial policy that derives from the mainstream Walras, Arrow, and Debreu (WAD) model.[16]

It is of importance to observe that the fundamental assumptions for WAD analysis are an assumed size and complexity of state space that makes it sufficiently transparent for central optimal choice and selection of winners to become a theoretical matter of intellectual simplicity, no competence, except the help of an economist, being required. It is a complete disaster to try it in practice, however. The experimental order and the distributed varied competence that I use for the state description of my policy analyses are the exact opposite to the assumed economic environment of the WAD model. This should be kept in mind in what follows.

The opening question for my analysis was the potential of biotechnology for industrial development. We have one problem of pure economic technological analysis. Is there sufficient competence in place to support the industry, notably customer and venture capital competence? A complementary policy problem is the peculiar ethical problems associated with biotechnology. The industry is said to "manipulate the deep mechanisms of life". Should this be allowed on ethical grounds? The problem is disturbing, considering the fact that biotechnology probably offers a great industrial potential for the advanced industrial economics. This has been discussed at length in Eliasson, Å. (2000) and I won't discuss it further here, except to observe that we have a very clear case of Coase (1960). You are harming me (read my sense of ethics) by carrying on biotechnology production. To that Coase would reply: by preventing me from carrying on production that you consider harmful to your moral ethics, you are harming me, and (I add) a lot, since if successful you may prevent the emergence of a new and viable industry. Now back to competence bloc based policy analysis and an illustrative example.

The Pharmacia reorganization into Pharmacia & Upjohn and the partial withdrawal of both R&D and production activities from the Uppsala region in Sweden pose an interesting practical industrial policy illustration with long run implications (see further Eliasson and Eliasson 2000); do we discuss the difference between the long run output of (1) the continued existence and growth of an established and advanced production of biotechnology and health care products or, (compared with) (2) its withdrawal from the region releasing locked up innovative resources for new industry establishment and growth?

Table 1 and competence bloc analysis (Table 3) give part of the explanation. Successful new establishment, notably university new establishment, requires the existence of a full-bodied competence bloc. Even with that influence we know that it will take time for newly established firms to gain momentum and sufficient volume to replace what has been withdrawn in the form of established volume production in Pharmacia. Pharmacia, however, only produced within an area, restricted in scope and variety to a particular set of technologies. The newly established firms based on previously locked up Pharmacia ideas exhibit much greater variety. In the very long run the probability should be higher that there will be more winners left and a larger production volume in the Pharmacia withdrawal case than the probability that production in a Pharmacia that would have remained would have been higher. This is, however a hypothetical prediction that is complicated by two circumstances. The first complication is that we are now discussing the innovative establishment of new firms. There is no back up "venture" financing of a large established firm. Hence, however large the technological and commercial potential, if one or two critical competencies in the competence bloc are missing, very little industrial will come about.

The second complication is that Pharmacia has released industrial competence (in the competence bloc, item 6) that is now supporting growth in the new firms. This makes the comparison somewhat unfair. Suppose instead [the third case (3)] that we have a complete green field operation in the form of new university establishments in the Uppsala region without support from released Pharmacia industrial competence. In this case, of course, the demands on a complete and full-bodied competence bloc is even larger. However, the middle scenario (2) also emphasizes another important industrial policy dimension, namely (Eliasson 1998c, Eliasson and Taymaz 1999) active policy support of the exit process to enforce the release of resources, notably competence, for new expansion in high value added production.

CONCLUSION

In general, the biotechnology and/or health care industry illustrates the fundamental competence problem of industrial policy (Eliasson 1998c). The early naive industrial policy debate of the 1970s was based on the central planning rationale of overview and the assumed possibility of picking the winners. Any form of competence beyond that overview on the part of the policymaker was ignored. And the economic advisers acquiesced. To be drastic, our analysis of the biotechnology industry demonstrates just the opposite; that of the importance of killing the losers fast as soon as they have been identified, to release competent resources for more experiments. However, for new industrial formation to succeed the competence bloc has to be complete to ensure that a maximum exposure to varied technological, entrepreneurial, financing and industrial competence takes place. This is needed to pull even a few winners out of a more or less unknown state space (business opportunity set). Even so the road to success will be fraught with business mistakes and society has to learn to live with a minimum rate of failure and also learn to admire those who dare to try, including those who fail. Here policy makers definitely have a role to play since they very much attempt to influence the value system in society.

Figure 1. The Health Care Industry

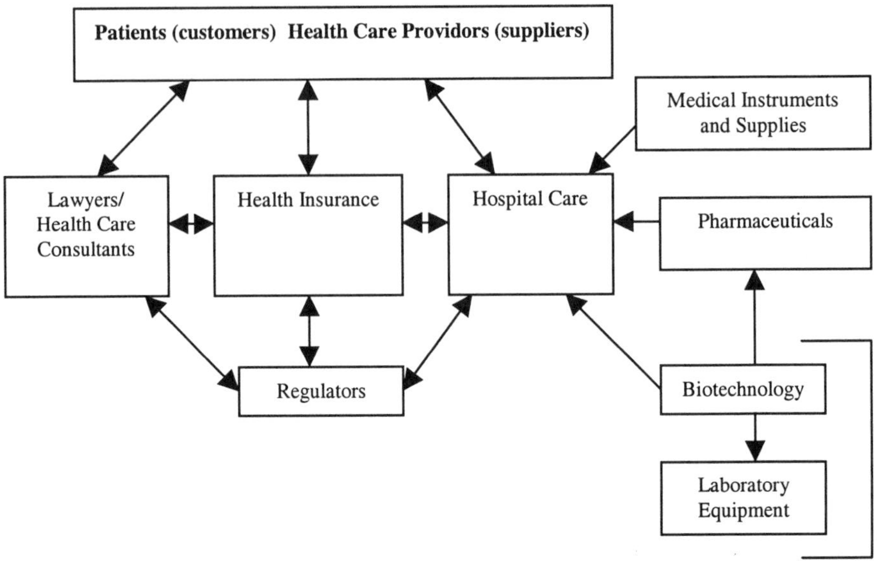

Patients (customers) Health Care Providors (suppliers)

Medical Instruments and Supplies

Lawyers/ Health Care Consultants

Health Insurance

Hospital Care

Pharmaceuticals

Regulators

Biotechnology

Laboratory Equipment

Source: Eliasson, Gunnar, 2000. The Health Care Competence Bloc – on the integration of bio, pharmaceutical and medical instrument technology with hospital care and health insurance into a new industry.

Chapter 4

BIOTECHNOLOGY:
Scientific Progress and Social Progress

Jacqueline Senker
SPRU, University of Sussex

No one can deny the rapid progress in scientific knowledge which has accompanied the explosion of government investments in biotechnology research over recent years. Scientific progress has been characterized by the rapid emergence of a succession of new techniques, such as PCR, receptorology, signal transduction, yacs, antisense and, more recently, bio-informatics, pharmacogenetics and pharmacogenomics.

It is no surprise that knowledge in the field has expanded so rapidly. Many governments have identified biotechnology research as a priority in their science and technology funding allocations during the last decade and they have also introduced measures to try and promote the application of that knowledge by industry. The intention is to use investments in science and technology as a lever for creating jobs and economic growth.

It is difficult to provide exact data of government expenditure on biotechnology research because there is no universally accepted definition for the field. Statistics are usually confined to broader categories such as life sciences. A recent EC BIOTECH project which prepared an inventory of biotechnology research in 17 European countries: Austria, Belgium, Denmark, Finland, France, Germany, Greece, Iceland, Ireland, Italy, Netherlands, Norway, Portugal, Spain, Sweden, Switzerland and UK (Enzing et al, 1999) attempted to identify how much money had been spent on public biotechnology research in the period 1994-98, and also to identify the fields to which that money had been allocated. The most difficult part of the exercise was developing a comprehensive taxonomy for biotechnology research. After prolonged debate the team agreed on the taxonomy, shown in Table 1, which includes medical research aimed at the creation of biological therapies (as opposed to small molecule drugs or research merely applying the tools and techniques of biotechnology).

Table 1. Taxonomy of Biotechnology Areas

B.1	Plant biotechnology (crops, trees, shrubs, etc.), including
1.1	reproduction and propagation
1.2	genetic modification introducing new/ excluding existing genes (mono- and polygenic traits)
1.3	growing conditions
1.4	plant protection
1.5	plant pathogen diagnosis
1.6	genome mapping
1.7	biodiversity of plants in agriculture/ horticulture
B.2	Animal biotechnology, including
2.1	reproduction
2.2	production
2.3	breeding, including genetic engineering in animals (creation of transgenics)
2.4	animal health care,
2.5	genome mapping
2.6	biodiversity of farm animals
B.3	Environmental biotechnology, including
3.1	microbial ecology
3.2	biosafety
3.3	microbial functions for degradation/ transformation of pollutants
3.4	isolation, breeding and genetic engineering of pollutants; degradation micro-organismes
3.5	biotechnological processes for soil and land treatment
3.6	biotechnological processes for water treatment
3.7	biotechnological processes for air and off-gas treatment
B.4	Industrial biotechnology: food/feed, paper, textile, and pharmaceutical and chemical production
4.1	enzymatic processes
4.2	development of bioprocessing techniques (fermentation, immobilisation of biocatalysts, quality control etc.)
4.3	downstream processing

B.5	Industrial biotechnology: Cell factory, including all biotechnology research focused on the cell as Producer of all sorts of (food and non-food) products
5.1	plant cell biotechnology: plant cell biology
5.2	animal cell biotechnology: animal cell biology
5.3	bacteria as cell factories: microbiology
5.4	genetic engineering and production of enzymes
5.5	genetic engineering of micro-organisms and yeast
5.6	cell culture techniques
5.7	genome mapping of specific bacterial and yeast genomes
5.8	biodiversity of micro-organisms in production processes
B.6	Developments of human/veterinary diagnostic, therapeutic systems
6.1	immunology, therapeutic and diagnostic antibodies
6.2	vaccinology
6.3	human genome mapping
6.4	human gene transfer techniques
6.5	therapeutic proteins and oligonucleotides (substitutes for pharmaceuticals)
6.6	tissue engineering
6.7	genomics in drug discovery (substitutes for pharmaceuticals)
6.8	DNA diagnostics
6.9	forensics (genetic fingerprinting)

B.7	Development of basic biotechnology
7.1	techniques to determine the structure of biomolecules and study the structure-function relationship.
7.2	techniques to build biomolecules (nanotechnologies)
7.3	interaction of biomolecules with micro-electronic devices, including biosensors, biomonitoring
7.4	genome analysis techniques
7.5	bio-data-informatics (tools applied to solve data handling and processing problems in biological research e.g. genome sequencing
7.6	bio-informatics (application of biological principles to information processing for technical applications)
B.8	**Non technical areas of biotechnology**
8.1	ethical issues
8.2	social issues (including public acceptance)
8.3	economic impact
8.4	ecological impact

Data was collected in each country from all the government departments, research funding agencies and charitable foundations which allocate funds to biotechnology research and found that our taxonomy was readily understood in all organisations and countries visited. The results of the inventory show that almost 10 billion Ecus was spent on biotechnology research in the period 1994-98. This amount includes

- expenditure on special biotechnology programmes;
- the biotech component of general technology programmes;
- the "open call" or response mode system of funding by research councils;
- block grants to research institutes;
- funding of government laboratories (institutes which carry out research for government ministries).

This total is certainly an under-estimate. Where possible we also tried to include regional funds, but this was not always possible. We also excluded local funding and the block grants to universities, usually provided by Ministries of Education which may also support biotechnology research.

Figure 1 shows how funds have been allocated between fields, with almost 50% being dedicated to human and veterinary biotechnology. There is limited availability of similar data from other parts of the world, except for the US. It was estimated that the Federal Government spent about $4.3 billion on biotechnology research in 1994 (Biotechnology Research Subcommittee, 1995) an increase of over 25% on 1990 expenditure of $3.4 billion (Office of Technology Assessment, 1991, 19). It is estimated that annual Federal investment in biotechnology research is now approaching $6 billion.[1] National expenditure is much higher than this because many of the states also fund research. There is no detailed breakdown of expenditure between areas, but it can be assumed that, as in Europe, that the lion's share of the funding goes to human and veterinary biotechnology; one indicator of this is the fact that medical sciences' share of all academic R&D has been rising since 1973 (the only field showing a consistent rise over time) and now accounts for 27.6% of total academic R&D in the US (Rapoport, 1998).

Figure 1. Funding of biotechnology areas in European countries

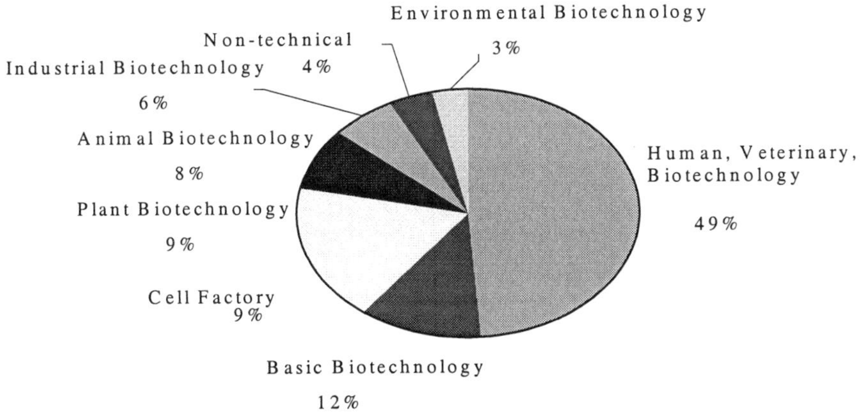

Governments invest in scientific research not only to achieve scientific progress, but in the hope of realising socio-economic benefits. General expectations in the early 1980s were that developments in biotechnology in the short term would be dominated by high value-added products, particularly those with medical applications. In the longer-term, agricultural applications were expected to have an even greater impact. At that time it had been widely accepted that genetic engineering research was not unsafe, but genetic engineering could present dangers in the future. Potential dangers, expected to relate to large-scale industrial production based on genetically engineered micro-organisms, demanded study and the implementation of appropriate regulation (Bull, Holt and Lilly, 1982).

How far have the expectations of the early 1980s about wider benefits of scientific progress been realised? The following sections deal with only three of the many broad applications for which biotechnology research promised to bring benefits: human health, animal health and nutrition and agriculture.

HUMAN HEALTH APPLICATIONS

With the highest biotechnology research expenditure focused on human health, it was anticipated that the first applications for biotechnology would lie in producing proteins and peptides for human therapy, or bio-pharmaceuticals. Other medical applications were predicted to be new antibiotics and vaccines, with new diagnostic techniques shifting the emphasis of medicine from treatment to prevention. It was thought that the impact of gene therapy, a method to treat genetic disorders, would be delayed because of controversial ethical issues surrounding the therapy (Davies, 1986). Many of the DNA or protein based therapies have not been as clinically effective as hoped, and many promising candidate drugs have failed at late stage clinical trials. Only 54 biotechnology derived therapeutics and vaccines had been approved for use in the US by the late 1990s, with an acceleration in approvals in 1998 (Morrison & Giovannetti, 1998). The majority of these drugs are for cancer and related conditions (17.1%) with infectious diseases, growth disorders and blood products each having more than 10% of the total. Drugs for AIDS/HIV infections and related disorders accounted for 8%. Somatic gene therapy involves inserting a functioning gene into the somatic cells of a patient to correct an inborn malfunction. Clinical trials in the US and Europe are currently evaluating the treatment of several genetic diseases as well as AIDS and various forms of cancer. But progress in reaching patients is likely to be slow because methods for inserting genes are still under development and not very reliable (Ronchi, 1996).

The main potential benefits to human healthcare have arisen from an area of research carried out on a global scale - the Human Genome Project - which did not emerge until the late 1980s. It aims to identify and map all the genes on the human genome. This project and the sequencing of other species have been specially favoured by funding agencies. It is estimated that since 1993 genome research around the world has received around $250M each year (Friedrich, 1996). The Human Genome Project hopes to publish a draft version (90%) of the human genome 18 months earlier than anticipated by February 2000. It has been asserted that the map will become a vital tool for medical research and could lead to the development of drugs and diagnostics designed for specific genetic traits (Chemistry & Industry, 1999). The Human Genome Project has also had wider impacts: on scientific progress itself, on controversies about the ownership of the knowledge and on concerns about the tests and therapies enabled by this knowledge.

Derek De Solla Price (1984) draws attention to the numerous occasions when the crafts and techniques which researchers develop to pursue their investigations, or what he called “instrumentalities”, have opened up major new opportunities for scientific investigation and innovation. The Human Genome Project has had just this effect. It has accelerated scientific progress through its invention of new experimental techniques devised for understanding and modifying genetic structures, the construction of databases of genetic information and by the development of complex computer systems, or bio-informatics, for exploring these databases. It has also led to the creation of the new sub-disciplines of genomics, proteomics, bio-informatics and pharmacogenomics. Small firms have been set up specifically to market their specialist expertise in these key-enabling technologies.

Scientific progress has also been accelerated by determination to complete the map before private sector rivals and so ensure that the complete human genome map remains in the public domain. Private companies plan to patent pieces of genetic information and make them available on a subscription basis only. Three patents have been issued to date by the US patent office and it is considering up to 10,000 other applications. There are hopes that negotiations between the UK and US will lead to agreement that all laboratories participating in the Human Genome Project will waive their patent rights so as to ensure that there is free access for researchers to the genomic information which may be used to develop diagnostic tests for predisposition to certain conditions - breast cancer, heart disease and diabetes - as well as therapies to treat them (BioNews, 032). Others, who recognise that disease has a multiplicity of causes, are less sanguine about the potential for genomic information to aid the development of new drugs. They argue that we do not understand the aggregate effects of the various genetic mutations and environmental factors which may dispose people towards certain diseases (Hodgson, 2000; Nightingale, 1999).

Nelkin (1995) has pointed out that there are numerous problems connected with diagnostic tests, including confidentiality of genetic data, or discrimination against people with specific genetic make-ups by insurance companies or employers. Diagnostic tests for certain conditions are progressing more rapidly than therapies to cure them, and create dilemmas for physicians. They may be called on to counsel healthy patients about future risks when no clinical solutions exist. There is also the possibility that "behavioural genetics" - predicting propensity to psychiatric disorders, addiction or aggression - may lead on to forms of social control.

> Diagnostic technologies are not intrinsically dangerous. Their significance rests on how they will be used. They call for educating

healthcare professionals about the possibilities and limits of DNA diagnostics, creating legislative provisions to prevent the discriminatory exploitation of genetic information.... and informing the public about the meaning and potential abuses of genetic information...". (Nelkin, 1995)

Expectations that research would lead to new vaccines and antibiotics have not yet been realised on a large scale (there are 3 vaccines in the 54 products with FDA approval). There is need for the production of new and improved low-cost vaccines, particularly for adults. New forms of vaccine are under research (naked DNA vaccines and cancer vaccines) but questions about their safety still remain open (Ronchi, 1996).

Social progress could be measured in terms of the extent to which drugs and vaccines are being developed and administered to all the world's peoples. A 1996 World Health Organisation-UNICEF report noted that unless the cost of vaccines drops to below $1 per dose, the vaccine is beyond the reach of much of the world. It also considers that "unless the international community continues to back scientific research and global immunization with adequate resources for new vaccines ... the great promise of molecular biology and genetic engineering may be squandered" (Fox, 1996).[2] Drug companies have developed expensive HIV treatments for rich western markets which extend the life-spans of many AIDS patients. Unfortunately sufferers in developing countries cannot afford to use these medicines. Some drug companies have reduced the price of AIDS drugs in some African localities. It has been suggested by Médecins sans Frontières that developing countries "should be able to obtain compulsory licences whereby national authorities allow local manufacturers to circumvent [drug companies'] patent rights, with certain conditions and in return for the payment of royalties to the inventor". African countries are reluctant to do this because they fear US trade sanctions (Borman, 1999). The World Bank is now seeking a partnership with pharmaceutical firms, backed by $5 billion, to invest in developing low-cost AIDS vaccines for the strains of virus and the health system capabilities of developing countries. The WHO and other international organisations will guarantee a market for these vaccines in developing countries (Elliott, 1999). Malaria affects 500 million people each year, killing up to 3 million. At least two teams of researchers, a US-Indian partnership, and a Colombian team are developing malaria vaccines but, even if successful, poverty may frustrate attempts to vaccinate the people in developing countries most at risk.

ANIMAL BIOTECHNOLOGY

Early predictions about potential applications of biotechnology to animal husbandry included the prevention of disease, stimulating growth and developing new animal types with predetermined characteristics (Davies, 1986). Over 120 diagnostic kits for pregnancy tests and disease were available by the mid 1980s (ACOST, 1990) and genetically engineered vaccines are available for several diseases (Houdebine, 1996).

Bovine growth promoters (BST) for dairy cows was another early introduction in some countries. BST has been approved and is widely used in the USA. There is extensive controversy about its use in Canada and its authorization is under review, but it is still banned in the EU. The basis for the EU moratorium was that there was no need to use it (overproduction of milk), rather than concerns about its safety. The EC now thinks that the safety of BST should be assessed and expert committees have given clear evidence of hazards to animals, and to humans (from drinking milk from treated cows) (Scientific Committee on Veterinary Measures Relating to Public Health, 1999). Optimism about the application of growth hormone has also been harmed by evidence of the effect on pigs. It gave them deformed bodies and skulls, swollen legs, ulcers, crossed eyes, renal disease and arthritis (Straughan, 1995).

Animal biotechnology has evolved in ways not foreseen by early commentators. The creation of Dolly the sheep has shown that animal cloning, the replication of entire animals, is possible. Cloning offers commercial opportunities when coupled with the production of transgenic animals – animals which have additional genes from other species inserted in their genomes. Transgenic animals have several potential applications, including making their organs more appropriate for transplant into humans (xenotransplantation), providing animals which act as better models for human disease or creating animals which express proteins in their milk which can be purified for pharmaceutical use, or can change the nutritional quality of the milk in some useful way. Three obstacles block the commercialisation of transgenic animals: the technical difficulty and high cost of the procedure (Moffat, 1998), and concerns that xeno-transplantation may spread pathogenic viruses from the donor species into human recipients. A recent study carried out by the US Centers for Disease Control and Prevention in conjunction with Novartis Pharma, however, found no evidence of pig viruses in 120 patients treated with living pig tissue (e.g. for grafts or pancreatic cells) (Paradis et al, 1999). The third, and perhaps major obstacle to the commercialisation of these techniques is the widespread controversy about their ethical justification. Discussion centres around the rights of animals, their welfare and the importance of maintaining respect for nature and the value of the natural world (for discussion see Straughan, 1995).

PLANT BIOTECHNOLOGY

A 1982 review of the potential of biotechnology noted that plant science was much less advanced than microbiology. It recommended support for research to increase basic knowledge of plant physiology and genetics, in order to allow biotechnology to be applied to agriculture. Noting that the supply of nitrogen was a major limiting factor to increasing plant production, it expected that biotechnology would have major impact on the nitrogen-fixing properties of important crop plants. It went on to suggest that food security for developing countries required crop diversity related to locally specific conditions, and that such diversity could be achieved by advances in biotechnology. Moreover, since the price of fertilisers was often beyond the means of developing countries, developments in improving nitrogen fixation could improve their crop (Bull, Holt & Lilly, 1982).

There has been rapid scientific progress in plant biotechnology even though it has received a small proportion of available public research funds (9% in Europe in the period 1994-98. See Figure 1 above). By 1990, genes conferring herbicide, virus and insect resistance had been isolated, characterized, reconstructed and introduced into crop plants. Environmental advantages were anticipated from a reduced use of pesticides as virus and insect resistant plants entered agriculture (ACOST, 1990). By 1995 crops had been genetically modified to make them more tolerant to salt and the knowledge provided by plant genome sequencing was expected to offer enormous potential for improving agriculture (Biotechnology Research Subcommittee, 1995).

The OECD's online Biotech Database (OECD, 1999) contains 55 genetically engineered crops which have been approved for commercialisation, have obtained approvals which allow commercialisation or are in the process of being approved. The majority (47) have gained approval for commercialisation in North America, with 31 of the applications related to herbicide tolerance; and a small number which incorporate pest and herbicide resistance. Another seven concern pest resistance, a few provide virus resistance or concern hybrid seed production.

Figure 2. Acreage Breakdown of ag-biotech crops

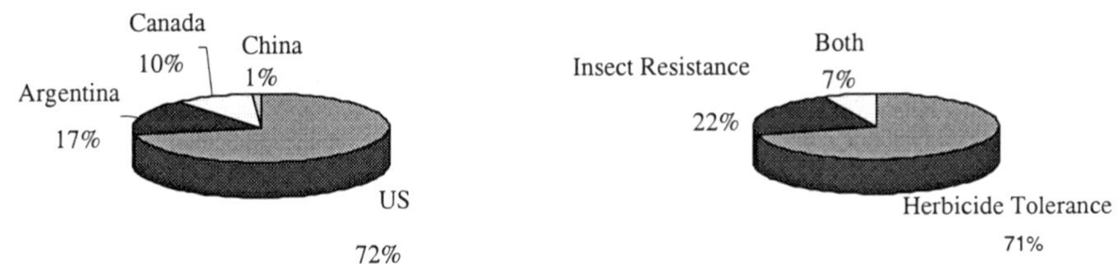

Acres planted with transgenic crops in 1999 – 98.6 million

Source: Chemical and Engineering News, 1 November 1999

In 1998 approximately 70% of field trials for GMO crops took place in the US and 15% in Europe (OECD1999a); 30% of US-planted soybean, corn and cotton were genetically engineered (Ernst & Young, 1999). There are a variety of reasons for the predominant activity in the US. These include the size of the market and the suitability to US climatic conditions of some of the crops first modified. Differences in regulation and in public acceptance between the US and Europe are more significant.

The US has a more relaxed and stable regulatory environment than Europe. Product approval comes under existing regulations which focus on the risks of the product per se. In addition, since 1993, there has been no requirement for field releases (for maize, cotton, potatoes, soybeans, tobacco and tomatoes) to be subject to an environmental evaluation. In contrast, the European system focuses on the process (genetic manipulation), and all innovative activity is subject to regulation. One of the reasons for introducing process-based regulation was to provide a harmonized regulatory framework in the EU. Thus the EU Directive 90/220 on the Deliberate Release of Genetically Modified Organisms (GMOs) stipulates EU-wide control on the deliberate release of all genetically modified organisms into the environment, whether for research in the laboratory, for experimental field trials or for commercial purposes. Directive 258/97 (the Novel Foods and Novel Food Ingredients Regulation) demands that all food derived from GMOs requires pre-market safety evaluation, and labelling for foods produced using genetic modification (for detailed information on EU and US regulation of genetic crops see Knight and Whitehead, 1994; House of Lords, 1998; FDA website).

A 1996 survey of public attitudes to biotechnology (Eurobarometer, 1997) found that the more useful biotechnology applications are considered to be for society, the more they are felt to be morally acceptable. Those interviewed considered that the application of biotechnology to food production and transgenic animals were the two applications posing the greatest risks to society. Between 1996 and 1998 UK opposition to the introduction of GM foods rose from 51 to 58% (New Scientist, 1998). Strong public opposition to agricultural biotechnology in Europe has been heightened by labelling requirements, and the decision of US agrochemical companies not to segregate genetically modified soya and maize in their exports to Europe. In response, many large European food manufacturers and retailers have decided to eliminate GM ingredients from the products they manufacture or sell. In addition to fears about the safety of GM crops, non-governmental organisations such as Greenpeace and Friends of the Earth have increased public concern by drawing attention to potential risks GM crops pose to the environment.

The evidence from Eurobarometer suggests that the public may be ready to accept GM crops which have clear social or health benefits. Examples, which come to mind, are pest-resistant crops, able to fix their own nitrogen which would reduce agro-chemical use and related water pollution; or crops tolerant to drought or brackish water which could greatly improve the life chances of those in sub-Saharan Africa. The public does not appear to discern any such benefits from current GM crops, and there is no evidence to date of commercial interest in crops which are tolerant to drought or brackish water. Moreover, a 1998 review (Cornerhouse, 1998) rebuts many of the claims for social benefits made by industrial proponents of genetic engineering in agriculture. It is encouraging to note that the Rockerfeller Foundation and other public agencies have supported research to improve the nutrient content of rice (iron and Vitamin A) and this could be significant to the health of people in developing countries where rice is the main food. The material will be freely available for non-commercial use in developing countries (Ye et al, 2000, Cockcroft, 1999)

DISCUSSION

There are many factors which explain the gap between early expectations about the application of biotechnology and what has been achieved. One important factor is scientific progress. Although it has been extremely rapid, there are still significant gaps in understanding which postpone the realisation of some goals. For instance, progress in developing nitrogen-fixing plants has been held back by the discovery that numerous genes control nitrogen fixation, rendering the modification of plants much more complex than if a single gene only was involved. This discussion, however, will focus on the equally important economic, social and political constraints which prevent the development and commercialisation of some socially acceptable and needed products.

Economic Constraints

Companies are driven by the need to make profits. In relation to plant biotechnology the major companies involved are oligopolistic, multinational agrochemicals companies. They have acquired seeds companies because for "genetic engineering of the higher plants to be privately profitable as marketed inputs, these commodities will have to be marketed as seeds" (Buttel et al, 1984: 341). Sylos-Labini (1969) has argued that oligopolistic competition holds back innovation, because "firms use their dominant positions in an industry to regulate the introduction of major innovations so as to minimise the cost of obsolescence." The nature of competition in this sector makes it highly unlikely that any agrochemicals company would take the risk of introducing seeds which might destroy its markets for pesticides and fertilisers. The early genetic engineering applications to plants by companies have, instead, focused on making them tolerant to the companies' own herbicides. The principal motivation for developing new products is to increase sales and profits, and, accordingly, companies are keen to develop product ranges which give them a competitive edge vis-à-vis competitors. The acquisition of seed firms by agrochemical companies and the development of crops which are resistant to their own pesticides (and in some cases unable to set seeds) appear to provide a captive market to these companies both for seeds for the transgenic crops and for pesticides. If they manage to secure a competitive edge over competitors, they can sell new products at premium prices, and they do this so as to recoup the costs of research and development.

The profit motive also explains why companies' concentrate on developing applications for wealthy markets in developed countries. They neglect developing cheap drugs or treatments for diseases such as malaria or AIDS which are devastating the lives of the impoverished in developing countries. There are no profits to be realised from countries which lack the means to purchase remedies.

Social Constraints

Some studies of innovation imply that there is an inner logic which determines the development and form of a new technology. This idea of "technological determinism" has been refuted by numerous writers who emphasise the social processes which influence the innovation process. Two main schools of thought exist to explain these social processes: the Social Construction of Technology (elaborated in Bijker, 1995) and Actor-Network Theory (Callon and Law, 1989). The former suggests that the social groups of organisations or individuals involved in the production and use of an artefact determine its design or form. Different groups attach a variety of meanings and problems to the same artefact, and different design solutions are proposed to meet different problems. An artefact reaches its final form when it ceases to be seen as problem and can meet the needs of all relevant social groups. For example, in the development of the bicycle the desire for speed by young men was solved by increasing the diameter of the front wheel of the penny-farthing. Safety for women cyclists led to a low front wheel and the introduction of inflated tyres. When it was shown that cycles with low front wheels and air tyres were safer and faster than penny-farthings, the penny-farthing died and the final form coalesced around the diamond frame which provides the basis for most contemporary bicycles. Actor-Network theory recognises that during the process of innovation scientists and engineers use a multiplicity of techniques and materials. They also engage in political, economic and social activities so as to enrol other actors and extend their influence beyond the laboratory. A technology is created when all these elements (which can include consumers, social movements, governments, companies and technical components) are brought together in a stable network relationship.

Both theories have important limitations, including their neglect of the cultural, intellectual or economic origins of the social choices about technology. They are important, however, in emphasising that there is a social dimension to choices about technology. These choices, however, are now influenced by a public which increasingly recognises the shortcomings of science and is unwilling to accept the reassurances of the scientific-political establishment (Beck, 1992). Beck suggests that these attitudes characterize "risk society", characterized by pervasive insecurity and a preoccupation with technological threats because of the failure to develop effective institutional controls or to recognise the limitations of science.

Political Dimensions

Regulation has been defined as state intervention in private spheres of activity to realise public purposes. Regulation traditionally had economic goals and sought to correct the activities of monopolies, trusts and cartels which led to high prices and shortage of goods In recent times regulation has also incorporated social goals, seeking to protect consumers from health and other risks (Francis, 1993). A great deal of current regulation is concerned with reducing risks.

> Science is employed at every stage in the regulatory process: ... But this is not to say that the scientific judgements incorporated in regulatory policy making are either clear-cut or beyond dispute. Many of the contemporary regulatory issues in which scientific judgements are invoked to assess risks are themselves the subject of considerable scientific dispute (Francis, 1993: 134).

There are disputes about why regulation occurs and whose interests it serves. Stigler (1971) argues that rather than regulation serving the public interest, it has been "captured" by industry and is designed and used primarily for industry's benefit and demands for regulation may be promoted by firms to secure or protect their markets (Peacock, 1984). Interest group theorists suggest that regulation is initiated by interest groups (e.g. Wilson, 1980) and regulatory outcomes are the result of interaction between the dominant group and the state, or compromises between groups. The public choice perspective shifts the focus of analysis from groups to the individual, and suggests that the competing preferences of individual actors can easily subvert the objectives of large organisations (McLean, 1987). A great deal of the regulatory literature is specific to certain contexts and there are difficulties in applying it to biotechnology. However the broad lesson to be drawn from this brief introduction to regulation is that political activities of those involved in regulation can affect its development and commercialisation.

CONCLUSIONS

Rapid scientific advance in biotechnology is a consequence of the large amounts of public money invested in the science base. Despite this scientific advance, early expectations about the very real social benefits that could be expected from these investments have not been realised. Moreover, some of the early innovations to come to market pose social and ethical problems. This does not imply that biotechnology of itself is either beneficial or harmful, it can be both. Companies' plans for the technology can only be realised to the extent that is allowed by social and political processes.

What is the role of scientists in this process? It is a role which can best be performed by scientists whose training incorporates social sciences, so they can better understand the economic, social and political factors which affect how scientific knowledge is used in society. Scientists' roles differ according to whether they are employed in the public sector or in industry. Industry's main objectives are to increase turnover and profitability, and this demands the protection of proprietary knowledge, including the work of its scientific staff. Industrial scientists are therefore limited in their ability to publicize any doubts they may have about company strategy for biotechnology, but they could give early warnings to company management about possible risks, or social and ethical attitudes.

The traditional culture of academic scientists gave them autonomy over decisions about the basic research to be performed and they aimed to publish and diffuse the knowledge gained. Autonomy and the freedom to publish are now coming under threat from government demands that academic research be relevant to economic and social needs and involve collaboration with industry. The inventory of European biotechnology research (Enzing et al, 1999) found that over 40% of total expenditure was "directed" in this way. The public count on academic scientists for information about the risks involved in new technologies, but their increased involvement with industry and/or reliance on industrial contracts may make such advice suspect. The potential to provide such advice is also limited by the lack of research on the environmental or health risks associated with genetic modification; such research only forms a small part of the 4% of European countries' expenditure on biotechnology research allocated to ethical and social issues or economic and ecological impacts. Increased attention by research funding agencies to bio-safety research could remedy this lack of knowledge.

Developing countries are not benefiting from developments in biotechnology because poverty renders their needs of little interest to profit-maximizing companies. International agencies are now collaborating with companies and charitable foundations to distribute some standard drugs and vaccines to developing countries. Social benefits from biotechnology could be realised if efforts were made to identify methods to promote the development and use of biotherapies for diseases endemic in developing countries. Similar actions to distribute seeds for nutritionally enhanced or salt-resistant crops to Third World countries have the potential to improve diet and health.

Crises such as global warming, and BSE, along with such disasters as Bhopal, Seveso, and Chernobyl, have raised public concern about the uses of technology. We now live in a risk society and the public is no longer willing to accept reassurances about safety from the scientific-political establishment. The potential for scientific progress in biotechnology to be translated into social progress requires both that scientists have the knowledge and environment to be open about the shortcomings of their scientific inquiries, and that the public is also fully involved in decisions about the regulation of biotechnology (Sagar et al, 2000). Public involvement could also help to build consensus about how to resolve the ethical and social issues connected with applications of biotechnology, and open the way for socially acceptable biotechnology innovation.

PART III

MEASUREMENT

Chapter 5

A CHALLENGE FOR MEASURING BIOTECHNOLOGY ACTIVITIES:

Providing a Comprehensive Perspective

Antoine Rose
Science, Innovation and Electronic Information Division
Statistics Canada

INTRODUCTION

While biotechnology is still at an early stage where a good part of its activities are based in research and development, various surveys conducted in Canada illustrated that biotechnology also has industrial activities that generate revenues. Therefore, not only is biotechnology a promising technology but it is actually diffusing and becoming an important social and economic part of the economy. A consequence of the production and diffusion of biotechnology is the rise of a number of policy challenges, which require statistical measurement in order to assist analysis.

BIOTECHNOLOGY DEFINITIONS

A major challenge in measuring biotechnology is to identify which activities are being performed by which actors. Once identification has been established, what are the proper groupings and which criteria should be used to classify the activities, the actors and the linkages?

For measurement purposes, biotechnology activities can be defined using various approaches, depending on the type of activity or respondent characteristics. The approaches could use single definitions, multi-stage definitions or a de facto definition derived from a list of biotechnologies, each approach with advantages and disadvantages. Two approaches used by Statistics Canada were the single definition and the de facto definition using a list.

Single Definitions

Several definitions of biotechnology already exist. The OECD definition, established in 1982, is: "the application of scientific and engineering principles to the processing of materials by biological agents to provide goods and services"(OECD 1982). A similar definition is used in Canada: "the application of science and engineering in the direct or indirect use of living organisms or parts or products of living organisms in their natural or modified forms". (Government of Canada 1999)

These definitions have been helpful to discuss what is meant by biotechnology and, what should be excluded or included or, what is modern versus old biotechnology. But, due to policy questioning on the importance of innovation in biotechnology, these definitions have been occasionally modified to add or subtract parts of sentences.

As an example: the following definition was used recently in the 1996 Statistics Canada questionnaire for the Research and Development in Canadian Industry Survey:

R&D in biotechnology is defined as the systematic investigation carried out in the natural and engineering sciences by means of experiment or analysis on biological agents in order to gain new knowledge and create new or significantly improved products or processes. «Biotechnology multidisciplinary nature encompasses a range of techniques dealing with recombinant DNA (Deoxyribonucleic Acid), cell fusion, plant and animal cell cloning, monoclonal antibodies, tissue culture, and bioprocess engineering. What distinguishes the new biotechnology from the one used to make bread, beer or cheese is the enormous expansion of our power to manipulate biological agents through the transfer modification, and control of genetic material. (Science Council of Canada 1989).

Three observations should be made about this definition:

- First, there is a clear wish to make a distinction between old and modern biotechnology and this is why there is an emphasis on words such as «systematic investigation» or the creation of «new knowledge». In this, the definition is consistent with the prescriptions of the Frascati Manual (OECD 1993).
- Second, biotechnology is not only a field of studies but also a set of techniques used for innovation purposes: creating new products and processes.
- Despite the fact that this seems to be a focussed definition, it would be surprising if a majority of respondents to the survey actually read this definition. As a consequence, biotechnology is somehow self-defined by the respondents.

Defining by Using a List

Rather than using a broad definition, the Biotechnology Use in Canadian Industry Survey conducted in 1997 by Statistics Canada used a list of 22 biotechnologies. The definitions for these 22 biotechnologies are based on more specific techniques such as recombinant DNA, peptide synthesis, bioremediation or tissue culture. A complete list can be found in Annex 1. The first criterion for being identified as an actor involved in biotechnology is: are any of these biotechnologies being used? Then, secondary questions such as the utilization stage or number of years in use can be asked.

The 22 biotechnologies cover a broad spectrum of techniques and were classified into three categories. The first, selection and modification of biological material, covers mainly the research and development stage where, biotechnologies are used to gather new knowledge. At this stage, biotechnologies are used to identify, select, extract, reproduce and amplify organic molecules such as DNA, peptide, antibodies and antigens.

The second category regroups biotechnologies, which are dealing with a higher macro-level of life material: the cell, tissue, bacteria, plants and animals. While these techniques may happen at a later stage after the initial selection/modification of organic molecules have bee made, it is not always the case and there is also research and development conducted. Still, the use of these biotechnologies is more frequently part of the production processes and therefore more on the use side of the knowledge.

Finally, the third category is a special case of biotechnology applications: environmental biotechnologies such as bioremediation. Although one may argue that most of these biotechnologies rely on the use of naturally occurring organisms, there is still work being done to improve their efficiency and scope of action.

Taken altogether, the list of 22 biotechnologies still constitute a fairly broad definition since it includes older techniques such as classical or traditional breeding of plants and animals. But, having responses for each of the 22 biotechnologies, different classifications of definitions using a subset can be tested.

This list of 22 biotechnologies was used in a survey that purpose was to establish adoption ratios, for the Canadian industry, on the use of biotechnologies. It provided descriptive information on specific use of biotechnology by different industries. Selection/modification biotechnologies are mainly used by the pharmaceutical industry and to some extent by the agri-food sector; culture/use of biotechnologies are mostly the domain of agri-food industries and; environmental biotechnologies are mostly used by industries related to natural resources such as forestry, pulp and paper, mining, oil and petroleum. (Rose 1998)

To answer different policy questions, another survey was designed, to establish characteristics of firms which are, to a large extent, dedicating themselves to the use of biotechnologies for developing new goods and processes. A new list of 17 biotechnologies was developed, largely similar to the first list. The classification was also changed. Environmental applications were subsumed into processing biotechnologies and a distinction was made for selection biotechnologies to distinguish between DNA related biotechnologies and biochemistry/immunochemistry based biotechnologies.

The virtue of these lists is to allow for direct measurement of biotechnology activities using a simple criterion: an actor is performing biotechnology activities if at least one biotechnology taken from an exhaustive list. Subsequently, the intensity of use can be measured to some extent by the number of different biotechnologies used; by the level of R&D performed in biotechnology or by a biotechnology derived sales to total sales ratio.

Statistics Canada used different lists of biotechnologies with some overlap in its two surveys, the first on biotechnology use in the industry and the second exploring firm characteristics. The second list has kept several biotechnologies listed in the first, slightly modified. Several lists, each serving a different purpose, are possible, but ultimately all the lists are likely to have similar features. This fact support the argument in favor of the development of internationally accepted master classification for biotechnologies.

In order to make international comparisons there is a need for a consensus on a list of biotechnologies. A classification system is helpful to better organize the list and to allow for several level of detail on the list. The result would be a list with several hierarchical levels, which would be used to survey biotechnology activities at a general detail level or at a more specific detail level.

Single Definitions vs Lists

In Canada, two surveys used a single definition of biotechnology (Industrial R&D Expenditures (Statistics Canada 1997) and Federal S&T Expenditures (Statistics Canada 1998) and two others used a list (Biotechnology Use by Canadian Industry and; Biotechnology Firm Characteristics (Rose 1998, BIOTECanada 1998). Use of a single definition is always somewhat problematic for an important reason: the interpretation of the definition is always left to the respondent.

First, the respondent must be knowledgeable about biotechnology. This is a necessary but not sufficient condition. Even the most knowledgeable respondent could not clearly identify what is included or excluded. Two examples illustrate this point. Several respondents in the Industrial R&D

Expenditures survey reported the performance of R&D activities in biotechnology, but the subsequent Biotechnology Firm Characteristics survey revealed that in fact biotechnology were not used. Follow-up discussion with the respondents confirmed that a single definition lead many respondents to respond positively to a question on performing biotechnology activities. When probed with a detailed question, using a list of biotechnologies, respondents excluded themselves.

The second example comes from recent testing of a new survey questionnaire on biotechnology activities in industry. The questionnaire included both approaches: a single definition up-front and a list of biotechnologies. Most respondents in a direct interview were puzzled by the single definition, to the point that, following common agreement with methodologists the single definition was removed from the final version of the questionnaire.

The Federal Science and Technology Expenditures Survey also used a single definition. This survey measures S&T activities of the Federal Government of Canada. The biotechnology definition and related questions were used to obtain the biotechnology share of overall S&T activities. This survey also used a different strategy by matching the respondent (usually from financial and operational side of departments) to a policy analyst specialized in biotechnology to help them decide on what should be included or excluded.

Other Definitions

Having defined a basic criterion for identifying an actor performing biotechnology activities and provided a mean to measure it, through a survey using a list, there is a need to have a dialogue on how the actors, the activities and all other variable such as how much, why, how connected and, what results ought to be presented and qualified. Having an actor performing biotechnology activities does not lead necessarily to defining biotechnology firms, a biotechnology industry or sector or, biotechnology products.

Perhaps biotechnology development will lead to the creation of new industries, as they are usually defined, in the future. In between, if one would like to study an ad hoc sector composed of firms, belonging to different industries, but having in common the fact that they are somehow dedicated to biotechnology, then, being a performer of biotechnology activities is a necessary condition but not a sufficient one.

One way to establish different criteria for biotechnology activities would be to start from products and processes and build on it. A first step would be first to define biotechnology products and processes the following way:

- Biotechnology product: a product which existence couldn't have been made possible without the use of biotechnology.
- Biotechnology process: a process for which the production of the output requires the use of biotechnology (here the use of present tense for the verb «requires» is essential. For instance, the production of beer is a fermentation process. Only if fermentation is defined as biotechnology would this process becomes a biotechnology. Assuming it may be argued that fermentation is not a biotechnology per se, then the use of genetically modified yeast leading to a significant improvement in the fermentation process instead of changing the end product characteristics would qualify as biotechnology process.)

A second step is to define a biotechnology firm as being engaged in the development of biotechnology products or processes. Already, a difficulty is being posed by the value of the engagement. Is this activity marginal or crucial to the firm? It may be argued that, given the requirement in Canada for an approval, by the government, of biotechnology products, being involved in the development of biotechnology products imply a resource investment which automatically qualify the firm as being a biotechnology one. The picture for biotechnology processes may not be as well defined since biotechnology may affect only a small portion of a complete process.

Assuming we now have biotechnology firms, it would be possible to construct different sectors by classifying them. Still, the aggregate of all biotechnology firms is too heterogeneous and needs to be refined by using classes based on end-applications such as agri-food or pharmaceuticals. Therefore, returning to standard industrial classifications. A biotechnology industry still does not exist. But at least there would be a method to identify, within existing industries, which firms are actually engaged in biotechnology.

COLLECTING INFORMATION

Although important, setting definitions is only the first step towards measuring biotechnology. The next step is to put these definitions through various data collection mechanisms such as existing surveys, new surveys or current administrative data collections. Beyond initial trials to measure biotechnology activities, the building of a comprehensive picture of biotechnology requires a more systematic measurement programme and some means for allowing comparative analysis with other countries.

Going back to the Framework for a Statistical Information System on Science and Technology developed by Statistics Canada (Statistics Canada 1998a), I will describe briefly for each item the correspondence for biotechnology measurement.

What?

For a large part, biotechnology is still in its development stage. Therefore, a research and development activity plays an important role. Beyond R&D, there are innovative activities, leading to intellectual property appropriation mechanisms, that is, products and processes being put to the marketplace where there are diffusion processes and adoption activities. But before going to market, biotechnology products and processes goes through regulatory processes, generating a different kind of activity: managing the various trial processes. Once on the market, products are sold, locally or abroad, generating revenue, exports, imports, profits…

The "what" question refers to the activities performed which, in the case of biotechnology, are the use of specific biotechnology, research and development activities, obtaining the approval activities, innovating activities and commercialization activities. Also, to be sure this list is not oriented only towards the industrial sector, we should add regulating activities and, education and training activities.

Who?

Actors can be found in three major sectors: higher education, government and industry. Research is being conducted by the higher education sector and by research institutes, several government departments are involved in research and regulation and several industry sectors are using biotechnologies. Beyond sectors, there are individuals, businesses, and organizations that are involved in biotechnologies. In the industrial sector, the real challenge is to find these actors since are they spread across a variety of other activities. For policy reasons, there is also a need to classify the industrial actors amongst biotechnology innovators or developers, adopters of biotechnology application and, services firms specialized in biotechnology.

Where?

The "where" question refers not only to geographical location but also to a sector dimension such as industrial classification. It is closely related to the question on actors. With the Statistics Canada Survey of Biotechnology Firms Characteristics (BIOTECHCanada 1998), an attempt has been made to use a more refined classification for biotechnology than the Standard Industrial Classification. This classification allowed for added distinction such as bio-informatics or between plant biotechnology and animal biotechnology in the agri-food sector.

Why and What Results?

The purpose for using biotechnology was dealt in Statistics Canada Survey on Biotechnology Use by Canadian Industry by following guidance provided by the Oslo Manual (OECD 1997) on the measurement the objectives of innovation. Questions were added on factors affecting the decision of firms to adopt biotechnology and, on results of biotechnology adoption. Items covered include costs, product range, and market position or production flexibility.

A uniqueness of the Biotechnology Firm Survey (1998) was a question on biotechnology related sales, which is also a direct measurement of results. Firms were asked to indicate what portion of their product sales was due to biotechnology products. On the whole, the results seemed consistent with other information gathered from the firms such as principal use of biotechnology and number of products on the market. Repeating the survey would give an idea of the growth. The particularity is that biotechnology related sales have been separated out from the revenue generated by the other activities of firms.

How Much?

This refers to the amount of resources committed to biotechnology. In Statistics Canada surveys, two type of measurement were done successfully: the first on dollars invested in the performance of biotechnology R&D activities and the second on personnel.

How Connected?

Following Oslo Manual guidance, there are questions on sources of information leading to the adoption of biotechnology. An alternative measure of «how connected» is to look for strategic alliances with other organizations (a strategic alliance being defined as a formal agreement with another firm to do business activities without merging).

Survey Methodology

This section describes the construction of the survey-population used in Canadian biotechnology surveys. For policy and analytical reasons, a result of the Canadian Biotechnology Use and Development Survey is to separate the survey population into three categories that can be analysed separately. First, respondents are divided between users and non-users of biotechnologies. Further down, it is also possible to divide between respondents that are only using biotechnologies in production and respondents that are using biotechnologies in an innovative mode to develop new products and processes. These two breakdowns may require different survey methodologies.

Making the first distinction between users or non-users of biotechnologies is done with a list of biotechnologies, where respondents are asked which biotechnologies are used. The criteria to be classified, as a biotechnology user is the use of at least one biotechnology from the list. The second distinction is made by using an innovation type question such as: "is your firm developing new products or processes that requires the use of biotechnologies".

Making a distinction between users and non-users requires that a representative population sample be randomly drawn for selected industries where there are reasonable expectations to observe the use of biotechnologies. Sample size is dependent upon the expected ratios of biotechnology users in the total population and desired level of data quality for the estimations. For instance, the Biotechnology Use in Canadian Industry Survey showed that 31% of firms are using biotechnologies in the pharmaceutical industry. It implies that responses from one respondent out of three can be used to characterize the biotechnology users.

Collecting information on biotechnology developers is more difficult. Apart from a few large firms, this population is composed in Canada of 75% of small and very small firms (under 50 employees). Several of these firms are early stage research entities that don't have revenue and survive by using venture capital. Capturing these entities requires constructing a take-all component in the sample. A take-all component that suppose the existence of a list of firms for which information permit one to expect their active involvement in biotechnology. In Canada, such a list is constructed using various sources: data from previous surveys, supplemented with the use of a commercially available directory.

The reporting unit for the surveys is the firm. The main reason for choosing firm over establishment as reporting unit is because the decision to use biotechnology for innovative purposes is a strategic decision made at the firm level. When a firm choose to adopt biotechnology, the corresponding capacity built can benefit to all the establishments of that firm.

However, there are two problems associated with the use of firm as the reporting unit. The first one is that some of the existing biotechnologies one could use are processes that are simply inserted into the production function. One example is the pollution control equipment used at the end of pipe used to clean wastes before they are discharged in the environment. This is a case where the responding unit might preferably be the establishment.

The second problem is related to the survey of large multi-establishment firms and that is not specific to biotechnology. For such firms, biotechnology activities may be conducted in a central research facility. Obtaining the information required to answer the survey questionnaire may involve several different persons in the firms, thus increasing the burden for the respondent.

CLASSIFICATIONS

Biotechnology activities are found in several industrial sectors. So, the existing industrial classification can be used to select specific industries for the purpose of a survey. For the time being, one can expect to find firms involved in biotechnology activities in industries using processes that transform materials originating from living organisms or using chemistry to transform other organic or inorganic material such as petroleum or minerals.

Biotechnology related products are usually related to health, food of the environment. These are all regulated domains that often require extensive testing for safety purposes and product approval processes. As a consequence, we can see several services firms involved in the development for various functions such as research, regulation process management. Software development or the gathering of venture capital. Therefore, there is also need for information on the service sectors. Again, the challenge is to find the firms involved in biotechnology development activities amongst the other firms engaged in similar activities.

Policy users as also expressed a needs for a different classification of industrial activities. Consequently, a classification was developed to collect information on the application domain of the biotechnology products and processes being developed. Examples of such domains that cannot be found in existing industrial classifications are health, human or animal, and applications of biotechnology for diagnostics, therapeutics or gene therapy.

CONCLUSION

This chapter has presented some ideas on how to put biotechnology definitions into use for the measurement of biotechnology activities which, in turn, serves to answer important policy questions posed on the development of biotechnology.

Biotechnology is a pervasive technology which is already impacting on a wide variety of aspects of our lives. In absence of solid and systematic measurement of biotechnology activities several policy questions are left unanswered. When answers exist, they may rely on partial facts or anecdotal evidence which have the potential to be misleading. Systematic measurement requires in turn research, analysis and the commitment to providing resources for an on-going exercise. It also needs consensus building on common and standard definitions and, ways and approaches to collect the relevant information.

It was explained in this chapter how a broad definition of biotechnology can be transformed into lists of measurable activities, linkages and outcomes. To begin with, a master list of biotechnologies needs to be established and agreed on. This list works as a filter or a basic criterion on which one could build definitions for other concepts such as biotechnology products, processes, firms or sector. Two lists of biotechnology activities were developed in Canada and used for direct measurement. Still, work need to be done to ensure the accuracy and completeness of a master list for international purposes.

Since biotechnology is now moving out of laboratories and into direct applications directly impacting on people and the economy also represent an opportunity for researchers specialized in science and technology issues to study, while it is happening, the diffusion of a pervasive technology and its impacts on society.

Chapter 6

MEASURING THE ECONOMIC IMPACTS OF BIOTECHNOLOGY:
From R&D to Applications

Anthony Arundel[1]
(MERIT), University of Limburg

INTRODUCTION

The consensus on the importance of biotechnology is backed by a series of studies over the past twenty years that have tracked the rapid growth of biotechnology research, as shown by the increase in the number of firms active in biotechnology (Morrison and Giovannetti, 1998), research collaborations (Lerner and Merges, 1998), patenting (Joly and de Looze 1996), and field trials of genetically modified organisms (GMOs).[2] Employment in dedicated biotechnology firms (DBFs) in the United States and Europe, as estimated by Ernst & Young, grew by 150% between 1995 and 1998 in Europe (Ernst and Young, 1999) and by 42% in the United States (Morrison and Giovannetti, 1998). The slower growth rate in the United States is due to a head start in the establishment of DBFs, with Europe currently catching up with the American lead. The number of GMO field trials in Europe, which is a measure of the application of genetic engineering to agricultural crops, increased from 1 in 1990 to a peak of 270 in 1997, although this was followed by a decline to 207 trials in 1998.[3]

These indicators on the number of DBFs, biotechnology patents, research collaborations, and field trials show that research in biotechnology has increased enormously since the mid 1970s. Yet, biotechnology is a still a long way off from reaching its economic potential. More than two decades after the first DBF was established in the United States, the estimated 1,283 American core biotechnology firms in 1998 were still unprofitable, spending 5.1 billion dollars more than their total revenues of 13.4 billion (Morrison and Giovannetti, 1998). The total 1998 revenues of all core biotechnology firms in the United States was less than that of Merck, the leading American pharmaceutical firm.

Stock market evaluations also point to problems in the development of biotechnology, in terms of the sector's ability to raise capital to finance further research and to cover its on-going losses. The stock value of biotechnology firms has gone through large swings in the late 1990s, sharply declining in 1998, with speculative capital flowing out of life science firms and into IT and Internet stocks. This was followed by a recovery in the value of biotechnology stocks in 1999, and then another sharp decline in early 2000.

IT is another generic technology with widespread applications. A comparison between biotechnology and IT highlights the limited economic success of biotechnology so far. Twenty-five years after the construction in 1946 of ENIAC, the first digital electronic computer, mainframe and mini computers were widely used in the United States and Canada for business applications in banking, payroll, and reservation systems (Lebow, 1995). In the late 1990s, twenty-five years after Cohen and Boyer's landmark discovery of how to insert new genetic material into a cell, biotechnology has produced a handful of valuable new drugs and vaccines, substantially improved versions of a few previously existing drugs such as insulin and human growth hormone, a controversial growth hormone for cows that is not permitted on the market in Europe and Canada; and corn, cotton, canola, and soybean varieties that have been genetically modified to resist proprietary herbicides or some insect pests. Although each of these products is of significance for health or for agriculture, their economic impact is far more limited than the impact of computerisation in 1970. In fact, even if all pharmaceutical products were based on biotechnology, the economic impact of biotechnology would be limited, since less than 2% of manufacturing employment in the United States and Europe is in this sector.[4] As noted by Burke and Thomas (1997), most of the economic impacts of biotechnology are likely to develop in other areas, such as in the agro-food sector.

The slow development of biotechnology, in comparison with IT, is partly due to unforeseen technical problems. This has slowed the application of gene therapy in medicine, while the use of advanced biotechnology to develop new crop varieties has been limited, so far, to the insertion of only one or two new genes per variety. Even then, the introduction of a gene for herbicide tolerance in soybeans could have resulted in slightly lower yields compared to conventional soybean varieties (Benbrook, 1999), which would reduce the economic benefits of genetically-modified soybeans, although other research provides more mixed results(Economic Research Service, 1999). The development of major technical advances in crop varieties, such as yields substantially above current levels (Ruttan, 1999), or cereals such as corn, rice and wheat that can fix nitrogen, have proven to be far more difficult than originally thought in the early 1980s.

These results show that there is a major gap between our economic expectations for biotechnology, based on the rapid increase in several measures of biotechnology research, and the actual economic impact of biotechnology, as shown by an evaluation of current economic indicators for sales and biotechnology employment, or a historical comparison between IT and biotechnology. The thesis of this article is that technical difficulties, although important, do not explain all of the gap between expectations and reality. Instead, part of the gap is due to the specific characteristics of biotechnology as a generic technology and the types of statistics that are gathered to evaluate the growth of the biotechnology 'sector'. These statistics, which focus on research-intensive activities, are unable to capture the extension of generic technologies from research to applications. The paradoxical result is that the available statistics on biotechnology both overestimate and underestimate the economic impact of biotechnology. The overestimate is due to assuming that the rapid growth rates in patents and research, particularly among pharmaceutical and agro-seed firms, can be directly applied to the economic effects of biotechnology. The underestimate is due to a lack of comprehensive data on the application of biotechnology outside of the main R&D networks.

Both an over- and an under-estimate of the economic effects of biotechnology are undesirable outcomes that need to be avoided. The economic impact of biotechnology is an important issue to policy makers and to business managers, both of whom require reliable data in order to develop appropriate policy and investment decisions. Furthermore, this data must be impartial and not biased in favour of venture capitalists, biotechnology firms, or government agencies with a mandate to encourage the development of biotechnology.

This chapter evaluates how the characteristics of a generic technology influences the types of statistics that are required to track its development and diffusion. The effect of the type of available statistics on our understanding of the economic impacts of biotechnology is then illustrated through three examples.

THE ECONOMICS OF GENERIC TECHNOLOGY

Biotechnology is a generic technology, with many potential applications, because of its characteristic as a process technology – it provides better ways of making many different things. At the same time, this means that biotechnology must compete directly with alternative process technologies. Under some conditions, these alternative technologies can complement biotechnology, as when conventional plant breeding methods are used to test the reliability of genetically-engineered plant varieties. Under other conditions, alternative technologies are direct competitors. The ability of alternative technologies to compete with advanced biotechnology is strengthened by the high cost to firms of acquiring biotechnology expertise (Arundel and Rose, 1999), technical problems leading to a long time to commercialisation, and, for some biotechnologies, the effect of public resistance on the market for products produced using biotechnology.

The pulp and paper industry provides a good example of the effect of alternative technology on the adoption of biotechnology. Biotechnology offers potential economic and environmental benefits at several stages in the manufacture of paper based on both virgin and recycled fibres. Pilot projects have established the technical viability of biotechnology to replace or complement existing chemical or mechanical methods for fibre pulping, pitch control, de-inking of recycled paper, bleaching, paper coating, and improving paper strength and drainage rates (OECD, 1998). Genetic engineering could also be used to develop improved types of fibres that require less expensive processing methods. However, most of the application of biotechnology to this sector appears to be limited to end-of-pipe environmental technology. Biotechnology still has only a minor impact on the manufacturing process for pulp and paper because of higher costs and reliability problems in comparison with existing technical alternatives.[5] Many of these problems will be overcome with additional research. But, the manufacturers of mechanical and chemical processes are not standing still – they are continuously improving the economic and environmental features of their own technology, which means that there is a continual increase in the standards that biotechnology-based processes will have to meet in order to be competitive.[6]

How does a technology develop in the face of competitive alternatives? Radical new technologies are usually more expensive than alternatives in the early years of their development because of long development times and a lack of economies of scale. Kemp et al (1999) provide evidence to show that radical technologies first develop in protected niches and only later diffuse to wider applications. These niches are most likely to develop for products for which there are no competitive alternatives. They are less likely to develop for process technologies, which by definition already have existing alternatives.

The pharmaceutical sector has probably acted as a protected niche for biotechnology because the market for pharmaceutical products is not particularly price sensitive. The price of a new drug that offers a solution to a previously untreatable condition can be set at very high levels. Furthermore, the buying public is prepared to pay substantially higher costs for a drug that offers minor or even negligible benefits over existing drugs for the same condition. This is partly due to market failure in pharmaceutical products, but it could also be due to market conditions that favour a 'love-of-variety', as shown in one economic model (Acharya and Ziesemer, 1996). These conditions increase the ability of pharmaceutical firms to recoup high initial investments to develop biotechnology expertise and products. In addition, there is strong public support for the use of biotechnology to develop new pharmaceutical products.

In contrast, the market for other biotechnology products, such as environmental technology, industrial process technology or agricultural seeds, is both price sensitive and competitive. Agricultural biotechnology is particularly interesting, since public resistance could further reduce the market price for genetically-modified crops in an intensely competitive market (Bijman, 1999). Under these conditions, the existence of alternative technologies will slow the application of biotechnology to agriculture and will invariably limit its application, over the short to medium term, to crops with very large markets. Alternative developmental methods, such as conventional plant breeding, can also be used as long as the genetic material for the desired trait can be found, or created by encouraging mutations, in the same species. As an example, herbicide tolerant corn was developed by Monsanto and AgrEvo using genetic engineering and by Cyanamid using other plant breeding techniques.

STATISTICS FOR THE ECONOMIC IMPACT OF BIOTECHNOLOGY

Statistics on biotechnology use should cover all firms that are potential users of biotechnology and evaluate the impact of alternative technologies on biotechnology use. The goal is to be able to answer a fundamental question: How pervasive is biotechnology? Or, what are the economic impacts of biotechnology outside of the sheltered market for pharmaceutical products?

These questions cannot be answered with the type of data that is commonly collected on biotechnology, nor by most economic research in this area. With a few exceptions, examined below, most of the economic data on biotechnology is for DBFs or for large pharmaceutical and agrochemical firms that perform biotechnology R&D. As an example, the popular Ernst and Young reports on biotechnology in the United States, Europe, Canada, and Australia are limited to 'core' biotechnology firms with a 'significant commitment of their overall business to biotechnology'.

Academic research on biotechnology similarly focuses on firms that perform biotechnology R&D, although the motivating factor is a fascination with innovation as a creative activity, and particularly with networking and collaboration between DBFs and large multinational pharmaceutical and agro-chemical firms. This focus can be readily identified by examining the topics covered in the economic literature, as indexed by Econlit, a major literature abstracting service of the American Economics Association that covers approximately 400 academic journals plus monographs. Econlit contains references to 169 publications on biotechnology between January 1990 and mid 1999.[7] An analysis of the titles and abstracts shows that the most frequent topic is collaboration and networking in biotechnology research, which accounts for 21% of the total number of articles, while an additional 10% concern intellectual property rights. Other major categories concern consumer opposition and the acquisition of biotechnology expertise in developing countries. The focus of the economic literature on research collaboration is certainly justified, given the well-documented history in the United States of new DBFs acting as gatekeepers between basic research conducted in universities and large established firms in the pharmaceutical sector (Acharya, Arundel and Orsenigo, 1998). However, very few studies evaluate the diffusion of biotechnology applications throughout the population of firms that are potential users of biotechnology.

What Types of Statistics are Needed?

Innovation researchers have long understood that the choice of statistics for measuring innovative activities can have a crucial influence on our understanding of innovation. This is one of the motivations behind the OECD's Oslo Manual, which develops indicators for innovative activities that are not based on formal R&D.

A recent study of the pulp and paper sector in Scandinavia illustrates how statistics can alter our understanding of innovation (Laestadius, 1998). The study finds that R&D data considerably underestimate innovation expenditures, since most of the developmental costs of new pulp and paper technology were not included in the firms' R&D accounts. The author of the study also argues that the adoption of biotechnology by firms in this sector will not be visible in standard Science and Technology statistics, which are more relevant for science-based, research intensive industries.

Similarly, it will be difficult to fully evaluate the economic importance of biotechnology if our knowledge is based on what we can see under a spotlight trained on firms active in biotechnology research, while everything else lies in semi-darkness. After twenty-five years of biotechnology research, the transformation of biotechnology from research to applications in sectors outside of pharmaceuticals and agro-seeds should be underway. New types of statistics are required to detect such a transformation. There are two main issues concerning the collection of new types of biotechnology statistics: how to define biotechnology and the types of statistics that are required.

Defining Biotechnology

Many chapters in this book articulate a variety of taxonomic definitions for biotech. Here I take, biotechnology to encompass a wide range of different technologies, ranging from first generation biotechnologies such as fermentation and plant breeding that have been in use for centuries to recent advances such as genetic engineering, protein synthesis, and cell tissue culture. The problem is that many studies of relevance to an evaluation of the economic impacts of biotechnology do not provide clear definitions of what is meant by 'biotechnology'. Biotechnology is sometimes limited to DNA-based biotechnology, sometimes to DNA-based plus other forms of advanced biotechnology such as protein sequencing and monoclonal antibodies, and sometimes biotechnology includes many different technologies based on the manipulation of biological materials.

Biotechnology statistics need to be based on clear definitions of what is and is not included as biotechnology. One issue is if biotechnology statistics should be limited to advanced biotechnology based on the direct manipulation of genetic material or if it should include other biotechnologies. The answer to this question partly depends on the purpose of collecting the data, but a wider definition, as long as the more advanced biotechnologies can be identified separately, is worthwhile. This is because many biotechnologies that are not based on genetic engineering today could incorporate genetically modified organisms tomorrow. An example is environmental biotechnologies, such as bio-remediation, bio-bleaching, and bio-augmentation, that are based on micro-organisms. There is no evidence that any of the applications of these technologies in Canada in the late 1990s used genetically-engineered micro-organisms. However, once environmental biotechnology is widely adopted, it should be a relatively easy step to use genetic engineering to improve the micro-organisms used in these processes (Arundel and Rose, 1999).

The need for clear definitions of biotechnology is also driven by the fact that there is no 'biotechnology' sector, which is partly, although not entirely, due to the generic characteristic of biotechnology. Industrial classification systems such as NACE in Europe, NAICS in North America, or the International Standard Industrial Classification (ISIC) system, do not contain a 'biotechnology' sector. Instead, biotechnologies are developed and used by firms active in many different industries. These firms also use and develop other technologies. This contrasts with IT, where new IT products are developed in the telecom, software and office equipment sectors, all of which are identifiable in industrial classification systems.

Data Requirements

As noted above, most available biotechnology statistics focus on biotechnology research, patents, DBFs, and alliances between DBFs and large firms. However, as the economic impact of biotechnology expands from research to applications, a different set of statistics are required in order to provide the type of data that is needed for an informed evaluation of the social and economic impacts of biotechnology. The same problem has been faced earlier in the type of statistics gathered for IT. There is now a plethora of statistics on IT use, such as the percentage of households with a PC, Internet use by households and by firms, and IT investment by firms across the full range of manufacturing and service sectors (OECD, 1999a). Furthermore, there is only a relatively short delay between the recognition of an application of IT, such as a business-to-business web commerce, and the development of statistics to track the phenomenon.

Table 1 provides a summary of the types of statistics that are currently collected on biotechnology and the types of additional statistics that are now required. The available statistics are suitable for an emergent technology, but current needs require statistics that are suitable for a technology in which applications become increasingly important. The statistics are divided into four categories using a framework developed by Statistics Canada (1998b): the generation of technology, its transmission to new users, how the technology is used, and its socio-economic impacts.

Table 1. Data Requirements for Understanding the Emerging and Applications Phase of "generic" technologies

Generation	Emerging Phase	Applications Phase
Transmission	• R&D • Venture capital • Public investment • 'Core' biotechnology firms	• Industry concentration • Adaptation of biotechnology to processes and products outside of core areas
Use	• Research collaboration • Expansion of ability to generate biotechnology	• Marketing collaboration • Expansion of ability to apply biotechnology
Socio-economic impacts	• Product niche markets where usual price-demand curves do not apply	• Extension to new markets where price-demand curves do apply

Statistics on the generation of emerging technology focus on the factors that help to create the technology, such as R&D expenditures, the availability of venture capital, public investment, and the number and types of firms that are capable of conducting research in the technology. However, as biotechnology develops into an applications phase, we need statistics on the number and types of firms that apply biotechnology to their industrial processes or manufacture products containing biotechnology. The types of statistics available for technology transmission needs to shift from a focus on transferring the ability to perform biotechnology R&D to the ability to apply biotechnology. Statistics on the use of biotechnology also need to include not only sheltered markets where normal price-demand functions do not apply, as in pharmaceuticals, but competitive markets where biotechnology must compete on price. This shift requires statistics on the use of biotechnology across all potential users within a sector. Finally, statistics on the socio-economic impacts of biotechnology need to expand to include new areas. Data on employment, for example, needs to include both the number of biotechnology employees in DBFs and the number of employees in firms that apply biotechnology as part of a range of other industrial activities.

The first step is to obtain data on the application of clearly-defined biotechnologies across entire sectors and the use of alternative technologies that compete with biotechnology. Several surveys in Canada and in Europe have moved in this direction. Two surveys by Statistics Canada on biotechnology, conducted in 1997 and 1998, asked respondent firms about their use of between 17 and 22 clearly defined biotechnologies.[8] The 1997 survey also obtained information on the application of biotechnology. A 1999 survey by MERIT in the Netherlands asked all firms in six European countries that produce agricultural seeds about their use of three different alternative methods for developing new crop varieties, including genetic engineering. The next section uses the results of each of these three surveys to illustrate the benefits of collecting statistics on biotechnology across entire sectors and on alternative technologies.

NEW STATISTICS FOR THE APPLICATION OF BIOTECHNOLOGY

Application of Biotechnology Across Sectors

The 1997 survey by Statistics Canada asked firms with over 5 million dollars in sales and active in one of 17 industrial sectors about the use of 22 different biotechnologies, divided into three groups: advanced biotechnology based on the selection and modification of biological material, environmental biotechnology, and bio-culture and industrial process biotechnology (Arundel, 1999). The advanced biotechnologies include the DNA-based technologies used in genetic engineering.

The 17 sectors were selected on the basis of their current and potential use of biotechnology. These included both sectors where biotechnology use was known to be highly likely on the basis of R&D activities (food, pharmaceuticals, and other chemicals), sectors where biotechnology was likely to be in use (mining, petroleum extraction and refining, and pulp and paper), and a mixed group of 'other' sectors where little was known about the actual use of biotechnology but where biotechnology has many possible applications (textiles, printing and publishing, instruments, and leather products). The survey asked about the actual application of biotechnology in addition to research. Responses were obtained from 2,298 firms, or 88% of the sample. An important limitation of the survey is that it does not include seed firms and many DBFs.[9]

The percentage of employee-weighted firms that used one or more of the biotechnologies in an application is given in Table 2.[10] The results provide an immediate picture of where advanced and other types of biotechnology are currently in use. The application of advanced biotechnology is almost entirely limited to the pharmaceutical and food sectors, while the use of environmental biotechnology is concentrated in resource-based industries. Very few firms in the 'other' manufacturing sectors where biotechnology has potential uses applied any of the 22 biotechnologies.

The results both indicate that the application of advanced biotechnology is highly constrained, and points to possible future areas of application. For instance, almost all of the environmental biotechnologies could be adapted, in the future, to use genetically-engineered micro-organisms.

Other information from this survey, and from the 1998 survey, show that the use of environmental biotechnology is growing more rapidly than the use of other biotechnologies. In addition, employment in firms that perform R&D in environmental biotechnology is also growing faster than in firms that develop other types of biotechnology, although the rapid growth is from a very small initial base of less than 500 employees (Arundel and Rose, 1999b).

Table 2. Percent of employee-weighted firms in Canada in 1996 that applied at least one technology from each biotechnology class (number of respondent firms in parentheses)

	Biotechnology class		
Sector	Advanced biotechnology[1]	Environmental biotechnology[2]	Bio-culture & industrial process[3]
Mining (49)	0	30	17
Petroleum, gas & coal (160)	0	51	11
Wood, pulp & paper (201)	< 1	68	13
Food & beverages (746)	13	18	22
Pharmaceuticals (61)	15	<1	38
Other chemicals (234)	3	12	3
Other manufacturing (559)	<1	2	3

Source: Arundel and Rose, 1999

1: rDNA, antibodies/antigens, peptide synthesis, rational drug design, monoclonal antibodies, gene probes, gene therapy, DNA amplification.

2: Bio-augmentation, bio-reactors, biological gas cleaning, bio-remediation, phyto-remediation.

3: Tissue culture, somatic embryo genesis, bio-pesticides, classical/traditional breeding, bio-processing, bio sensing, bio-leaching, bio-bleaching, microbio-inoculants.

Biotechnology Employment

Government support for biotechnology is partly based on the expectation that biotechnology will create not only new employment, but better paid, highly skilled employment. Employment benefits are also stressed by biotechnology trade associations. An extreme example is a report for EuropaBio, which estimates 3.65 million biotechnology 'dependent' jobs in Europe in 2005, given favourable regulatory conditions (EuropaBio, 1997), although this estimate is based on several untenable assumptions about what constitutes a biotechnology 'dependent' job. An evaluation of the study methodology shows that almost all of these jobs are due to reassigning existing jobs in the agro-food sector to biotechnology.

Estimates of the amount of new employment created by biotechnology is fraught with problems, since part of the new employment will be balanced by job losses in other sectors or among employees using different technologies. An additional problem is that very simple estimates of employment in DBFs are also subject to errors. For instance, it is unlikely that all of the 153,000 American and 46,000 European employees in DBFs, based on the Ernst and Young estimates, were actually involved in or dependent on biotechnology in any way. The reason for this is that the number of jobs linked to biotechnology declines as firms grow in size and diversify into different technologies or businesses.

Both the 1997 and 1998 Statistics Canada biotechnology surveys provide relevant data on this point. The 1997 survey asked the manufacturing and resource sector firms for the number of their employees that 'worked with biotechnology' in 1996. The 1998 survey results are limited to 191 core biotechnology firms that perform R&D. Each respondent to the 1998 survey was asked to estimate the number of their firm's employees 'engaged in biotechnology activities' in 1997. These included seven different types of jobs that were related, in any way, to biotechnology: R&D, clinical affairs/quality assurance, regulatory and related affairs, manufacturing, marketing and sales, business development/finance, and administration/human resources. The definition of biotechnology in both surveys includes both advanced DNA-based biotechnology and other types that are not based on genetic engineering. The average percentage of employees in each survey with 'biotechnology-dependent' jobs is given in Table 3. The results are limited to firms that either apply biotechnology or perform biotechnology R&D.

Table 3. Average percentage of employees in biotechnology-using firms that are involved with biotechnology[1]

1997 survey of manufacturing & resource sector firms			1998 survey of core R&D perfoming biotechnology firms		
Number of employees	N	% biotechnology employees	Number of employees	N	% biotechnology employees
< 50	29	36%	< 10	49	88%
50 - 99	24	27%	10 - 49	84	86%
100 - 499	49	7%	50 - 99	20	71%
> 500	58	2%	> 100	38	35%

[1]: The 1997 survey results are limited to firms that use biotechnology and which replied to the question on the number of employees that 'work with biotechnology' (160 or 60% of 271 firms). The 1998 survey results are based on the number of employees, in seven activities, that are 'engaged in biotechnology activities'. 191 of 199 eligible firms (96%) answered the question.
Source: Statistics Canada, Survey of Biotechnology Use in Canadian Industries - 1996 and Biotechnology Firm Survey - 1997. Analyses by the author.

The results of the 1998 survey are the most comparable to the Ernst and Young estimates since they are limited to firms that perform biotechnology R&D. In small firms with less than 50 employees, most of which are DBFs, an average of over 86% of employees are involved with biotechnology in some way. The percentage of biotechnology employees remains high, at 71%, for firms with between 50 and 99 employees, but drops sharply to 35% for firms with over 100 employees. Once all employees are summed across all of these firms, there are an estimated 6,940 biotechnology employees, with is equivalent to 29% of all employees among these firms.

These results show that estimates of biotechnology employment that are based on counting all employees in core biotechnology firms as in biotechnology jobs will overestimate real biotechnology employment. The degree of overestimation will increase with firm size.

Most available estimates of biotechnology employment only look at firms that perform biotechnology R&D and ignore the left-hand side of Table 3 - or biotechnology employment among all resource and manufacturing firms. These firms show an even greater drop in biotechnology employment by firm size, with only 2% of the employees in large firms with more than 500 employees having jobs that involve biotechnology. However, an analysis of the 1997 survey results produces an estimate of 5,520 biotechnology employees in the resource and manufacturing sectors, of which 35% are in the food and beverage sector. After excluding double counting of firms that were food and beverage sector. After excluding double counting of firms that were covered in both surveys, the best estimate is that there were about 11,500 biotechnology jobs in Canada in 1997. One-third of these jobs were in manufacturing and resource firms that primarily use biotechnology in applications. Most of these jobs are not identified in surveys that focus on firms that perform biotechnology R&D, leading to an underestimate of biotechnology employment. This example shows that all potential users of biotechnology need to be surveyed to provide an accurate estimate of biotechnology employment.

Alternative Technologies and Sector Effects

The next example, based on agricultural biotechnology, illustrates the important role of alternatives to DNA-based biotechnology. It also provides another example of the advantages of surveying all firms active in a specific sector, rather than only firms that use advanced biotechnology.

The use of genetic engineering to develop new varieties of agricultural crops is the second most economically important application of biotechnology to date, after pharmaceuticals. All of the main multinational seed firms have research programmes to develop genetically-modified seeds for major crops such as corn, cotton and soybeans in the United States and sugar beet and oilseeds in Europe. An evaluation of the statements made by major seed firms such as Monsanto or Novartis suggests that genetic engineering is the most important technology for developing new crop varieties. Buttel has recently concluded that there are still some "modest pockets in which non-biotechnology approaches (such as traditional plant breeding) still predominate"...but "in mainstream agricultural research circles across the globe biotechnology or genetic engineering is largely the accepted approach" (Buttel, 1999). This interpretation of the current dominance of genetic engineering is backed by a common belief that only genetic engineering can substantially increase current yields, either through improving biological efficiency or by reducing crop losses through greater pest resistance, better tolerance of adverse conditions such as saline soils or drought, or improved storage attributes (Prakash, 1999).

It could be true that genetic engineering is the future of agriculture, but a more cautionary approach to the current importance of genetic engineering is warranted on the basis of statistics that are not limited to the main multinational seed firms. An analysis of the European data on GMO field trials shows that 74% of all trials are for a very limited range of crops: corn, sugar beet, and oilseeds (mostly canola).[11] Furthermore, genetic engineering faces competition from several alternative methods of developing new crop varieties.

The effect of alternative technologies is illustrated through the results of a MERIT survey in 1999 of European seed firms in six European countries: the UK, the Netherlands, France, Germany, Denmark, and Spain. The survey sampled all seed firms, including foreign subsidiaries, active in the development of new agricultural crops. The response rate was 72%. Results were obtained from 99 firms that developed new agricultural plant varieties.

Each respondent was asked to estimate the percentage of their firm's crop development budget that was spent on three different technologies: genetic engineering, assisted conventional plant breeding, and conventional plant breeding alone. The 'assisted' category is a hybrid between genetic engineering and conventional methods. It uses gene sequencing and DNA markers to speed up the development of new plant varieties. The distribution of development budgets in 1999 and the expected distribution in 2002, weighted by the number of employees to account for differences in the size of crop development budgets, is given in Table 4.

Table 4. Distribution of current and expected crop development budgets by 99 European seed firms (Employment weighted)

Developmental technology	1999	2002
Genetic engineering	10.2%	14.7%
Assisted conventional breeding[1]	16.5%	26.2%
Conventional plant breeding	73.4%	59.1%
	100%	100%

1: Conventional plant breeding assisted by DNA markers, gene sequencing, etc.

The results are in sharp contrast to Buttel's estimate that genetic engineering dominates crop development methods, with only 10.2% of the crop development budget spent on this technology. Alternatives to genetic engineering dominate the development budget. However, the share of conventional plant breeding is unlikely to reflect its share of expenditures to develop new plant varieties, since some of the expenditures on conventional plant breeding will be used to test the reliability of varieties that have been developed using genetic engineering. In contrast, almost all of the spending on assisted conventional breeding will be for new plant varieties. A maximum estimate of the share of genetic engineering in the development of new plant varieties can be obtained by assuming that only genetic engineering and assisted conventional techniques are used to develop new varieties. On this basis, a maximum of 38% of the 1999 budget for new crop varieties would be for genetic engineering [10.2/(10.2 + 16.5)*100]. This is expected to decrease slightly to 36% in 2002 because of the more rapid increase in the use of assisted conventional breeding compared to genetic engineering. Of course, the actual share of the development budget due to genetic engineering will be less than these estimates because many varieties, particularly crops with relatively small markets, are still developed through conventional breeding technology.

The share of the development budget that is due to genetic engineering is most likely higher in the United States and Canada, since the use of genetic engineering in Europe is probably hampered by public opposition to GMO foods and uncertain regulatory conditions. Nevertheless, the development of new crop varieties takes between seven and ten years, which means that it is difficult for European firms to alter their development projects in the face of current uncertainty. Furthermore, interviews with European seed firms show that they do not expect public opposition and regulatory uncertainty to last in the long term.

Even if European seed firms use genetic engineering less than American seed firms, the results of the European survey convincingly show that genetic engineering does not yet dominate the development of new crop varieties. Alternative technologies are still widely used.

CONCLUSIONS

Over the past two decades, there have been intermittent periods of optimism and pessimism over the economic impact of biotechnology, partly due to unexpected technical difficulties and public opposition, and partly due to incomplete statistics that focus on biotechnology research, giving a misleading picture of the economic importance of biotechnology. In comparison with other generic technologies such as IT, biotechnology has had a more difficult time in breaking out of the laboratory and into product and process applications. One result is that predictions for the economic impact of biotechnology have been moving from the present or very near future into the more distant future. A good example of the latter is the foreword to a 1999 Ernst and Young report on Australian Biotechnology by Nick Minchin, the Minister for Industry, Sciences and Resources for Australia. Mr. Minchin strikes a comparatively cautious tone, stating that "the biotechnology industry will become a key driving force in economic growth and employment over the next 30 years" (Ernst and Young, 1999b).

Although Mr. Minchin's assessment of a long time line is probably more accurate than statements that imply that biotechnology has 'arrived' today, biotechnology has been shifting from an emerging technology in which most economic activity is concentrated in research to a technology that is being applied to production processes and to the manufacture of new products. This development is apparent in the 1997 Statistics Canada survey on the use of biotechnology in several manufacturing and resource-based sectors. An increase in the application of biotechnology requires new types of statistics that can provide accurate estimates of the economic effects of biotechnology, such as biotechnology employment. There are three main requirements: the collection of data from all firms in sectors where biotechnology has current or potential applications, clear definitions of what is meant by 'biotechnology', and the collection of data on alternatives to biotechnology. All of these requirements need to be met in order to provide an unbiased appraisal of the economic impact of biotechnology.

The three examples given above illustrate how the economic impacts of biotechnology can be over or underestimated when these requirements are not met. The first example shows how data on the application of biotechnology by firms that do not necessarily perform biotechnology R&D provides more accurate information on biotechnology applications than when biotechnology statistics are limited to core biotechnology firms. This example also highlights the value of clear definitions of what is meant by 'biotechnology'. The second example on biotechnology employment shows that simple estimates based on total employment in core biotechnology firms will overestimate the number of actual biotechnology jobs. Conversely, a lack of data on biotechnology use by all firms in sectors where biotechnology has potential applications will underestimate biotechnology employment. The third example on the types of available technologies for developing new seed varieties points to the dangers of assuming the predominance of biotechnology when alternative technologies are available. This example also shows that all firms active in a sector need to be surveyed in order to estimate the prevalence of biotechnology use.

Finally, this chapter stresses the collection of statistics to evaluate the economic impacts of biotechnology. However, other socio-economic statistics, such as on environmental impacts and benefits, quality of life measures, and outputs, will also increase in importance as biotechnology develops from an emerging to an applied technology. These social and environmental effects of biotechnology will probably vastly outweigh the actual economic effects of biotechnology, as measured in terms of employment or sales. Adequate socio-economic statistics will be increasingly important in the future, although at this time the predominant need is for better economic data.

ACKNOWLEDGEMENTS

The analyses of the two Statistics Canada surveys on biotechnology were funded by the Science and Technology Redesign Project of Statistics Canada, currently SIEID. The survey of European seed firms was funded by the TSER Programme of the European Commission under Project PL 97/1280 (PITA). All information, opinions, and errors in this article are the sole responsibility of the author only and cannot be attributed to either of the funding agencies.

Chapter 7

STRATEGY AND PERFORMANCE FACTORS BEHIND RAPID GROWTH IN CANADIAN BIOTECHNOLOGY FIRMS

Jorge Niosi
Cirano, University of Quebec at Montreal

INTRODUCTION: EXPLAINING THE GROWTH OF THE FIRMS

Evolutionary economics and management has documented significant and stable differences in the performance and behaviour of firms operating in the same markets and using similar technologies. In other words, they found large differences in performance (growth rates, profitability, market shares) among competitors operating in similar industries and using similar technological paradigms.

At the basis of such wide differences they have found several factors. In the first place, there are different resource endowments. For instance, some biotechnology firms have been founded by, or have enrolled, "star scientists" whose knowledge and prestige are used as levers for attracting venture capital and other investments (Zucker and Darby, 1999). Other firms have been unable to attract them or even to detect their importance. In the words of a leading analyst of strategy "the limited information-processing capabilities of top managers give rise to very real barriers to efficient growth."(Hill, 1995: 319) Secondly, managers have divergent views of appropriate goals, structure and strategies for firms. Thus, some firms have pursued biotechnology applications in agriculture, others in environmental bio-remediation, in food products and in human therapeutics and diagnostics. Also, managers vary widely in their disposition towards risk and uncertainty. Thirdly, some firms may possess unique attributes, such as a superior proprietary technology that can yield better results (Nelson, 1995). Finally, some firms display superior organizational capabilities or routines allowing them to conduct R&D or to manufacture products in a more efficient or effective way than their competitors.

Both evolutionary theories and strategy analysts emphasize firm differences and are prepared to find out and understand differential and stable patterns of performance, such as different growth rates, or different levels of profitability, and relate them to different resource endowments, organizational routines or managerial capabilities and goals. Also, evolutionary theorists have underlined the fact that due to initial investments in human and non-human capital, due to sunk costs and established contracts, technological and organizational trajectories, firms experience major difficulties in radically changing their development paths after having started operations. In the management literature this dimension is called "organizational inertia". In other words, once the emerging firm has chosen its production niches and markets, they will not easily be modified. This characteristic applies to biotechnology firms that operate in wide different niches such as human health, food, ag-bio and environment. Once they have chosen their niche, and made irreversible investments, they will not easily change routes and will most often prosper or die with their initial goals and assets (Ryan et al., 1995).

CANADIAN BIOTECHNOLOGY

Biotechnology

The first commercial biotechnology firm, Genentech, was founded in 1976 in the United States, and went public in 1980 (McKelvey, 1996). Twenty-four years later there were some 1400 dedicated biotechnology firms in the United States, and some 4200 in the world. Human health was the area in which nearly half of world firms were operating. In this sector, as well as in plant biotechnology, small emergent biotechnology firms tended to grow in geographically localized clusters (Swan, Prevezer and Stout, 1998; Niosi and Bas, 2000). Also, they showed a marked propensity to organize alliances with research universities, government laboratories and other firms for cooperative research and development (R&D). They also cooperated with large international firms with the aim of conducting R&D, particularly clinical essays, but also for manufacturing and marketing their products (Arora and Gambardella, 1990; Barley et al., 1992; Dodgson, 1993; Pisano, 1991; Powell et al, 1996; Barbanti et al, 1999). Most of these dedicated biotechnology enterprises grew in association with venture capital firms that provided funds and managerial expertise to the knowledge-intensive start-ups (Kenney, 1986; Orsenigo, 1989). However, no author has precisely measured the importance of alliances on the growth of the DBFs, the role of alliances as a factor of their competitive performance.

Canadian Biotechnology

Canada's first biotechnology firm was probably Connaught, originally founded in 1914 by Nobel Prize winner Frederick Banting to produce vaccines, and later commercialize the new processes to produce insulin. By 1997, Canadian biotechnology was composed by some three hundred dedicated biotechnology firms (DBFs), of which some seventy-one were quoted in the stock exchanges in June 1999. Nearly 46 per cent of these three hundred firms were operating in the area of human health products, 22per cent in ag-bio, 11 per cent in food products, 7 per cent in environment and 14 per cent in other areas (Government of Canada, 1999). By mid-1999, the seventy-one corporations quoted in the stock exchanges had a total market capitalization of C$13 billion, but three of them represented over two-thirds of that amount (see Table 1, at the end of this chapter). These were QLT in Vancouver, Biochem Pharma in Montreal and Biovail in Toronto. It is to be noted in Table 1 that not a single Canadian company among the largest by market capitalization is operating in any sector other than human health. Also, in the 1990s, the entire Canadian biotechnology community had received over 3 billion dollars in investments (including venture capital, private placements, initial and subsequent public offerings). Human health firms had collected over 95 per cent of that amount (Tables 2 and 3, at the end of this chapter). In other words, the Canadian investment community had turned its back to all fields of biotechnology other than the pharmaceutical. Also, by mid-1999 only 62 Canadian firms had obtained patents in the United States, and a few others had obtained Canadian patents. In other words, most Canadian DBFs had no exclusive intellectual property of their own, in an area where intellectual property is seen as a major attractor of financing.
(Please refer to tables 1 to 3 at the end of this chapter)

These new Canadian bio-pharmaceutical firms showed a similar propensity to conduct alliances in R&D with both local institutions and large foreign firms. Their propensity to conduct cooperative research was close to one (Niosi, 1995; Walsh et al, 1995).

TWO SETS OF DATA

In order to find out the determinants of growth of the Canadian biotechnology firms, I built two sets of data. The first set came from a survey

I conducted during the summer of 1999, of sixty Canadian biotechnology firms. The second was a database I built in the Winter 2000 using several sources of quantitative information. Both sets of data yielded similar results.

The Survey

Early in 1999, I derived a list of DBFs using the Canadian Biotechnology Directories (Contact Canada, 1993 to 1998) published between 1993 and 1998. I completed the preliminary data set with Canadian Biotech News, a specialized journal yielding a few more names (Canadian Biotech News Service, several issues). These directories and journal provided information about the number of total employees, total sales and total revenues of these firms. I found that less than fifty firms (out of over three hundred existing in 1999) were experiencing rapid growth. Rapid growth was defined as 50 per cent growth in either sales or employment between 1993 and 1997, the last year for which information was available in the spring 1999. Only firms with sales or revenue over C$2 million, or over 25 employees were included in this category. Most firms had experienced little growth or no growth at all. I built a random sample of thirty rapid-growing firms and another sample of similar size of companies that were not growing or showed slow growth. During the summer of 1999, I interviewed the sixty firms personally using a questionnaire of some 110 variables previously tested and tuned up with the help of Statistics Canada's Science and Technology Redesign Project.

A few factors explained most of the growth of the firms. Growing firms were at least ten years old, they aimed at the production of diagnostic and/or pharmaceutical products, they had obtained patents and venture capital, they exported their products and they had conducted strategic alliances, often with foreign corporations.

Almost no firm outside human health had experienced rapid growth. The typical stagnant firm was operating in ag-bio, food production or environmental protection niches, had not requested patents, had not obtained venture capital, was not quoted in the stock exchanges, and did not export its products. Besides the lack of access to capital, its major obstacle to growth many of these firms encountered consumers' acceptance problems. Foreign markets were increasingly reluctant to accept genetically modified plants or animals. Local markets were resistant to bio-remediation of all sorts, thus condemning environmental firms to starvation, or forcing them to jump into export markets. Human health firms did not encounter such types of consumers' resistance, and attracted – as we have seen – almost all venture capital and financing.

Exploiting the human health niches was a necessary condition but not a sufficient one. Some human health firms were opposed to patents because they feared not to be able to defend their intellectual property against counterfeiters; others found that patents were too costly, or that they released too much sensitive information. However, companies without patents experienced difficulties in attracting venture capital. Private placements and business angels did not replace the management knowledge that venture capitalists could bring to human health firms, but could represent helpful additional sources of capital for the venture-capital backed firm. Conversely, not all firms with patents attracted venture capital, as this indicator of novelty was not the only basis on which venture capital firms allocated their funds.

Lacking experience in financing, some of these firms without venture capital support were launched into the stock markets at the wrong moment, or without appropriate announcements and communications, and they obtained only a few million dollars in their initial public offerings (IPOs). These amounts were often too small to finance the development of the human health products that were in their pipeline. Also, DBFs were often unable to launch further stock offerings in the market.

Patents usually helped to obtain venture capital, but some DBFs with other assets had been successful in attracting this type of investment. Thus a few firms without patents, but having enrolled star scientists, or targeting very promising market niches, managed to get financed by venture capitalists.

Alliances were not a sufficient condition of growth either for human health DBFs. Some of them had signed alliances too early in their product development process, when the value of their intellectual property was yet difficult to ascertain. After the alliance was signed, these DBFs found that their larger partners did not pay sufficient attention to them or did not contribute to the development of their products. Also, most local alliances were aimed at research and development, while most foreign partnerships, more relevant for the cash flow of the DBFs, were aimed at manufacturing and marketing products, and they were conducted with large overseas pharmaceutical corporations. Foreign alliances arrived thus later in the life of the DBF. Few companies had managed to sign this type of alliances.

In other words, each and all the factors behind rapid growth had a good statistical relationship with the dependent variable, but none had a very strong one (see Table 4 at the end of this chapter).

However, the logistic regressions using these variables allow us to explain most of the growth of the DBFs. Each logistic regression beats randomness by over 30 per cent, and explains most growth (see Tables 5 and 6 at the end of this chapter)

A New Sample

On the basis of the Canadian Biotechnology Directories, completed with additional information from the Canadian Biotech News, I built two other samples in order to retest the significant statistical relationships that the survey had found. I used these Directories to find out the firms' rate of growth, and to build new samples of soaring and stagnant enterprises. The new database was completed with patent data from the US Patent Database, and information about venture capital obtained from different public sources.

In the new random samples, there were forty fast-growing firms and forty stagnant ones. In the first group, thirty-one DBFs were active in human health, four in ag-bio, four in other specialties (mostly environment and bio-informatics) and one in food products. Their average age was 18.5 years and the median was 11.5 years. In 1997, they had an average of 82 employees and a median of 39. Half of them had been granted patents, almost invariable from the United States Patent Office.

The comparison was made against another random sample of forty companies with no growth or slow growth. The average age of these stagnant firms was 14.8 years and the median age was 11 years. Sixteen of these firms were in human health, while 24 of them were in other areas, most often ag-bio and environmental bio-remediation.

The results of the comparison between growing and stagnant firms yielded similar results compared to the survey. A few variables explained most of the growth. These variables were age, patents, venture capital, foreign alliances, exports and human health. Growing firms were older, had obtained patents and venture capital, targeted export markets and operated in human health niches. Most interesting three variables explained most of the growth: these were human health, foreign alliances and exports (See tables 7 and 8 at the end of this chapter).

CONCLUSIONS

This study has shown that very different rates of growth experienced by biotechnology firms, can be explained by a few variables, that are somehow related to strategy: choice of product niches (human health against all others), patenting strategy, financial and organizational decisions. The growing firms were active in human health niches such as diagnostics, contract research, bio-medical devices, but above all therapeutics. They also requested and most often obtained patents, had been financed by venture capitalist (and often attained the stock markets with their help), and targeted international markets with the support of overseas partners, most often large multinational pharmaceutical corporations.

Some level of organizational inertia was also evident. Basically no firm had radically changed niches, while a few ones operated in two different niches (such as neutraceuticals and human therapeutics, or human and animal health). As suggested by evolutionary analysis, DBFs tend to live and die with their initial choices. Strategic decisions take place at the beginnings, during the launching period. Once irreversible investments and contracts have been made, firms are usually unable to change directions.

The study thus suggests that future firms would have to make key decisions at the start, including the choice of the right niches, and that operating firms need to pay attention to their patenting strategies, their financial structures and sources, and their marketing choices. Timing and preparing the alliances, as well as the IPO, seem crucial: not all alliances and financing are equally related to growth and success. The earlier the alliance, the more difficult it will be for the foreign partner to evaluate the intellectual property of the DBF. Conversely, the later the partnership the greater the risk of the DBF running into cash flows problems. This is why alliances are not always related to success. DBFs need to weight the relative merits and difficulties of the timing of the alliance.

Also, stock markets tend to appreciate the quantity and quality of the information released by the firms before and after the IPO. Entering into the capital market is costly because of the permanent flow of information that these markets require in order to maintain the price of the DBFs shares, allowing it to go into future public offerings.

Table 1: The Largest Canadian biotechnology firms by market capitalization (As of Sept. 4, 1999, millions of Canadian dollars)

Company	Sector	Metropolitan Area	Capital. (C$M)
Biochem Pharma	Therapeutics	Montreal	3839
QLT Phototerapeutics	Therapeutics	Vancouver	3450
Biovail	Therapeutics	Toronto	2100
Patheon	CRO	Toronto	391
Forbes Meditech	Therapeutics	Vancouver	375
Visible Genetics	Genomics	Toronto	276
Cangene	Therapeutics	Toronto	273
World Heart	Bio-medical devices	Ottawa, Ont.	235
Biomira	Therapeutics	Edmonton	204
Dimethaid Research	Therapeutics	Toronto	159
Hemosol	Therapeutics	Toronto	153
Anormed	Therapeutics	Vancouver	140
Angiotech Pharmaceuticals	Therapeutics	Vancouver	133
Aeterna Laboratories	Therapeutics	Quebec, Que.	133
Technilab	Therapeutics	Montreal	128
Axcan Pharma	Therapeutics	Montreal	126
Bioniche Life Sciences	Therapeutics	London, Ont.	109
Nymox Pharmaceuticals	Diagnostics	Montreal	100
Vasogen	Therapeutics	Toronto	97
Synsorb Biotech	Therapeutics	Calgary	87
Draxis Health	Therapeutics	Toronto	79
Inex Pharmaceuticals	Therapeutics	Vancouver	77

Sources: Canadian Biotech News, Vol. 8, No. 36, September 11, 1999, p. 6.
Contact Canada: Canadian Biotechnology 1999 Directory, Ottawa, 1999.

Table 2: Bio-investments by source

	Venture Capital	IPO	Private placement	2 PO, public offerings	Other	Total
1991-1997						
C$M	555.69	566.29	449.94	821.16	58.03	2451.11
%	22.7	23.1	18.4	33.5	2.3	100
Number of placements	148	32	65	34	13	292
1998						
C$M	275.49	5	97.5	9.7	160.36	548.05
%	50	1	18	2	29	100
Number of placements	72	2	10	2	8	94
1999 (6 months)						
C$M	129.35	87.31	84.12	217.90	0.5	519.18
%	25	17	16	42	0.1	100
Number of placements	51	6	10	5	1	73

Source: National Research Council of Canada

Table 3: Bio-investments by field

	Human	**Agriculture food**	**Fish, forest environment**	**Other**	**Totals**
1991-97					
C$M	2327.26	96.94	26.91	-	2451.11
%	95	4	1	0	100
Number of placements	263	22	7	0	292
1998					
C$M	514.52	22.64	10.90	-	548.05
%	94	4	2	0	100
Number of placements	84	7	3	-	94
1999 (6 months)					
C$M	503.77	3.9	11.51	-	519.18
%	97	1	2	0	100
Number of placements	67	3	3	-	73

Source: National Research Council of Canada

Table 4: Survey Correlation (Pearson)

	Ragro	Age	Hhealth	Patent	Vencap	Alliance	Forall	Delays	Public	Consac1
Ragro	1.000									
Age	.149	1.00								
Hhealth	.242	.252*	1.000							
Patent	.235	.106	.337 **	1.000						
Vencap	.052	.249*	.223	.381 **	1.000					
Alliance	.330 **	.065	.094	.136	.114	1.000				
Forall	.185	.321 **	.264	.218	.061	.797 **	1.000			
Delays	.139	.064	.117	.043	.017	-.068	-.061	1.000		
Public	.328 **	.006	.410 **	.322 **	.100	.148	.182	-.171	1.000	
Consac1	.042	.354**	.055	.070	.072	.092	.044	.072	.153	1.000
Export	.237	.179	.146	.088	.036	.131	.233	.003	.223	.122

** Correlation significant at 0.01 level (2-tailed)

* Correlation significant at 0.05 level (2-tailed)

Table 5: Explaining rapid growth

Variables entered on step number

V 1	AGE (Number of years after foundation)
V 5.1	HHEALTH (Products targeted for human health)
V 7.1	EXPORT (Company exports products, Y/N)
V 13.1	PATENT (Number of company patents)
V 14.1	VENCAP (Company obtained venture capital, Y/N)
V 18	ALLIANCE (Company conducts alliances, Y/N
V25.16	CONSAC1 (Consumers' acceptance is major problem for those not growing)
V3.6 RAGRO	(Rapid growth), dependent variable

-2 Log Likelihood	51,018
Goodness of fit	84,783
Cox & Snell	0,307
Nagelkerke	0,412

	Chi-square	Df	Significance
Model	18.719	7	0.0091
Block	18.719	7	0.0091
Step	18.719	7	0.0091

Classification table for V3.6
The cut value is 0.50

Observed	Predicted Y	Predicted N	
Growth Rapid	16	6	72.73%
Not rapid	4	25	86.21%
Overall	20	31	80.39%

Variables in the equation

Variable	B	S.E.	Wald	Df	Sig	R	Exp. (B)
V1:AGE	.0509	.0371	1.8804	1	.1703	.0000	1.0522
V5.1: HHEALTH	.9358	.7331	1.6293	1	.2018	.0000	2.5492
V7.1: EXPORT	1.4961	.9528	2.4657	1	.1164	.0817	4.4643
V13.1:PATENT	.1088	.0739	2.1660	1	.1411	.0488	1.1149
V14.1:VENCAP	1.4614	.8197	3.1789	1	.0746	-.1300	.2319

V18:ALLIANCE	3.9976	37.183 7	.0116	1	.9144	.0000	8.2303
V25.16.CONSAC 1	–1.5100	1.5371	.9650	1	.3259	.0000	.2209
Constant	-4.1881	1.5712	7.1053	1	.0077		

Table 6: A second logistic regression on growth

Variables entered on step number

V 1:	AGE (Number of years after foundation)
V 7.1	EXPORT (Company exports products, Y/N)
V 13.1	PATENT (Company has patents, Y/N)
V 14.1	VENCAP (Company obtained venture capital, Y/N)
V 19.7	FORALL (Company conducts alliances with foreign partners, Y/N)
V25.16 CONSAC1	(Consumers' acceptance is major problem for those not growing)
V3.6 RAGRO	(Rapid growth) dependent variable

-2 Log Likelihood	29.517
Goodness of fit	29.622
Cox & Snell	0.484
Nagelkerke	0.646

	Chi-square	Df	Significance
Model	27.102	6	0.0001
Block	27.102	6	0.0001
Step	27.102	6	0.0001

Classification table for V3.6
The cut value is 0.50

Observed Growth	Predicted Y	Predicted N	
Rapid	15	4	78.95%
Not rapid	3	19	86.36%
Overall			82.93%

Variables in the equation

Variable	B	S.E.	Wald	DF	Sig	R	Exp. (B)
V1:AGE	.1589	.1124	1.9975	1	.1576	.0000	1.1722
V7.1: EXPORT	3.4605	1.4639	5.5882	1	.0181	.2517	31.8341
V13.1:PATENT	.2376	.0955	6.1942	1	.0128	.2722	1.2682
V14.1:VENCAP	-3.6421	1.5759	5.3412	1	.0208	-.2429	.0262
V19.7: FORALL	3.6939	1.5809	5.4599	1	.0195	.2472	40.2030
V25.16 CONSAC1	-4.3018	2.2104	3.7876	1	.0516	-.1777	.0135
Constant	-6.6240	2.5800	6.5917	1	.0102		

Table 7: Pearson Correlation. Second Database

	Age	Hhealth	Exports	Patents	Vencap	Locall	Forall
1. Age	1.000						
5.1 Hhealth	.062	1.000					
7.1 Exports	.158	.102	1.000				
13.1 Patents	.181	.439**	. 194	1.000			
14.1 Vencap	.193	.359**	.066	.437**.	1.000		
18. Locall	.112	.189	.059	.190	.148	1.000	
19. 7 Forall	.016	.011	.116	.471**.	.394**	.289**	1.000

** Correlation significant at 0.01 level (2-tailed)

* Correlation significant at 0.05 level (2-tailed)

Table 8: A final, more economical, logistic regression about growth

Variables entered on step number

V. 19.7 FORALL (Y/N)
14.1 VENCAP (Y/N)
5.1 HHEALTH (Y/N)
V. 3.6 RAGRO (Y/N)

Estimation terminated at iteration number 4 because
Log likelihood decreased by less than 0.1 percent

-2 Log Likelihood	69.799
Goodness of fit	70.451
Cox and Snell – R^2	.354
Nagelkerke – R^2	.473

	Chi square	df	Significance
Model	32.299	3	.0000
Block	32.299	3	.0000
Step	32.299	3	.0000

Classification table for RAGRO
The cut value is .50

	Predicted Yes	Predicted No	Percent correct
Observed			
Yes	26	8	76.47%
No	9	31	77.50%
Overall			77.03%

Variables in the equation

Variable	B	S.E.	Wald	df	Sig.	R	Exp. (B)
Forall	1.6409	.5994	7.4947	1	.0062	.2320	5.1597
Vencap	1.7242	.6246	7.6213	1	.0058	.2346	5.6079
Hhealth	.9844	.6157	2.5560	1	.1099	.0738	2.6762
Constant	-2.2574	.5896	14.6593	1	.0001		

Chapter 8

ASSESSING THE ROLE OF THE UNIVERSITY OF CALIFORNIA IN THE STATE'S BIOTECHNOLOGY ECONOMY

Cherisa Yarkin
University of California at Berkeley

INTRODUCTION

The United States is at the forefront of a worldwide transition to a "new economy" of knowledge-driven economic development. Biotechnology, information and communications technologies have revolutionized the way products are developed, spurring the growth of entirely new industrial sectors. One area of particular policy interest is the role of academic research and training in creating S&T-based economic benefits. An increasing proportion of S&T research is being conducted at universities,[1] but economists are still in the formative stages of developing a complete understanding of how these investments generate economic returns.[2] A wide array of models, definitions, and datasets has been developed, but there remain gaps in our understanding of the channels through which benefits are created.[3]

The University of California (UC) Critical Linkages Project aims to advance understanding of the role of public investments in academic science, at a time university-based fundamental research and advanced education is seen to be of increasing relevance to regional, national, and world economies. The Project explores new ways to assess the contributions made to the California economy by public R&D investment (through support for basic research and education) over the past twenty years. It focuses on the individuals who participate in that research and education as primary sources of contributions that help determine the course of economic growth. This approach complements and extends traditional assessment methodologies by moving beyond standard measures of publications and patents, and focusing on the people who generate the new knowledge these proxy measures are meant to reflect. It further enriches the standard view by identifying the multiple and varying roles that knowledge generators play over time.

The UC Biotechnology Research and Education Program launched the Critical Linkages Project in 1995 with the initial charge to develop a case study focused on the California economy and a particularly robust area of regional economic development: commercial biotechnology. The Project undertook to identify, document and develop a methodology for assessing and quantifying those contributions that accrue from publicly funded basic research and graduate education and from the relationships that develop between UC scientists and commercial biotechnology firms in California.

The subset of biotechnology activity that we examine in this case study comprises a substantial proportion of total commercial activity in the biotechnology sector. California is home to more than one third of U.S. public biotechnology companies, which in 1996 accounted for more than 56% of industry research and development expenditures, and generated 63% of total US commercial biotechnology revenues (Lee and Burrill, 1996). Commercial biotechnology in California provided more than 48,000 jobs that year, with average annual salaries of $60,000. These patterns have remained strong in subsequent years, despite the rapid pace of change in the sector.[4]

The initial findings of the Critical Linkages Project biotechnology case study illustrate the essential role publicly funded research and education have played in the development of commercial biotechnology in California, which can be viewed as the field's center of origin and a continuing source of economic growth and innovation.

THE CRITICAL LINKAGES PROJECT

The Critical Linkages Project was launched in 1995, at a time when the US Congress was undertaking a highly skeptical review of public investments in scientific research and graduate education. Research universities, including the University of California, were challenged to provide tangible evidence of economic benefits accruing to academic research and graduate education. This charge posed a challenge to economists, as well, because of the difficulty in making strong economic claims based on the traditional academic research productivity measures of patents and publications.[5]

We have taken the view that publicly funded research and education is best viewed in the context of the broad range of contributions universities make to economic growth. As an initial step in contributing to this discussion, the Critical Linkages Project launched a case study of the relationships between research and education at the University of California, and commercial biotechnology in the State. In the spirit of the Zucker et al,[6] and

Audretsch and Stephan (1996), we focus on the activities of individuals. For biotechnology, these contributions include scientific leadership in founding or substantially contributing to the establishment and success of biotechnology firms; graduate education in the highly competitive and creative basic bioscience research environment, which prepares the highly skilled workforce needed to successfully develop, produce and market new biotechnology based products; and scientific innovation which produces discoveries that advance fields of inquiry into exciting and often unexpected directions, new research findings that form the substrate for licensing and future commercial trajectories, or fruitful early stage research collaborations with companies that create a confluence of research strengths from universities and industry to advance important new frontiers, such as genomics, or to establish proof of concept of nascent discoveries.

We wish to emphasize the preliminary nature of this study, noting that a great deal of work remains to be done. For example, it would be of interest to examine the effectiveness of alternative technology transfer policies, the relationship between local economic development efforts and geographically localized industry clusters, the value of graduate education to different strata of biotech firms, and the extent to which ongoing relationships with academic researchers contribute to business success.

Methodology

The initial assessment of commercial biotechnology activity in California required that we define ‘biotechnology’ and determine which business entities constituted the appropriate target of study. For the purposes of this study we focus on the subset of firms for which biotechnology is a primary activity.[7] We adopt the definitions of ‘new biotechnology enterprises’ introduced by Zucker, Darby and Brewer (1994).[8] Our definition is consistent with that found in Eliasson (1996), and includes recombinant DNA technology, the use of antibodies (often termed cell fusion), and protein engineering.[9] For the purposes of this study, a ‘core’ biotechnology company is one in which the tools of molecular biology are used in research, development and, where applicable, production of a product. We count separately each business establishment that (a) has a recognizable, separable identity, and (b) undertakes biotechnology-based R&D activity. Following Zucker, Darby and Brewer (1994), this definition includes both start-up companies (what they term ‘new biotechnology enterprises’) and subunits of existing firms (their ‘new biotech subunits’).

These definitions had important implications for data collection. Major sources of economic data are the databases maintained by government agencies, such as the US Patent and Trademark Office and the Bureau of the Census, and proprietary databases of patent citations and journal articles. These sources were unsuitable for the initial definition of the population of biotechnology companies in the State for several reasons. Government databases, and many proprietary information services, categorize and report data using standard industrial classification (SIC) codes. These codes use a hierarchical system of increasing specificity to define industry sectors and organize data. For new technologies and emerging industries, the SIC codes do not reflect the categories that actually define the industry. For example, some biotechnology companies that develop therapeutics are categorized under SIC code 2834, Pharmaceutical Preparations, along with many companies that use standard biochemistry. Other biotechnology companies are classified under SIC code 2836, Biological Products, but companies that simply produce plasma and serums are also categorized under this code. A number of organizations that report on biotechnology activity, such as the California Healthcare Institute, do not separate biotechnology companies from biomedical companies, while others rely on company self-reporting instead of a particular definition. Therefore, in order to construct a complete, well-defined dataset, we used a multi-step approach.

Firms were identified using industry directories, and through consultation with experts at the University and in the private sector. The initial list of "core" companies was developed by undertaking a systematic review of the following biotechnology industry directories: *BIOSCAN Directory of Biotech Companies* (1994 and 1995); *Coombs Biotechnology Directory* (1995); *Standard and Poor's Corporation Register of Directors and Executives* (1995); *The Bioscience Directory, San Diego County Edition* (1995); and *California Biotechnology Corporate Directory* (1995). The initial list was vetted with a broad range of experts, including a biotechnology analyst at Ernst and Young LLP; Mark Edwards, Managing Director of Recombinant Capital, a San Francisco-based consulting firm specializing in biotechnology alliances and capitalization; the 12 UC Biotechnology Research and Education Program Executive and Advisory Committee members; several other UC faculty members who were also company founders; UC campus alumni association and development offices; and UC patent coordinators.

Approximately 400 California companies were assessed for inclusion in the database. By applying a definition of 'biotechnology' that emphasizes the use of modern molecular techniques, and after accounting for firms that had merged, moved out of state or gone out of business, the list was narrowed to 228 firms. It should be noted that because the number of California biotechnology companies is continually changing, the CLP "core" list is regularly updated to reflect the most current available information. Although this paper reports only our initial findings from the original data set, the database is structured to track information about each company over time, so that UC contributions can be traced as the industry evolves.

Once the dataset was constructed, we assessed the opportunities to use existing databases, such as those maintained by federal, state and local agencies with responsibilities to track business data (e.g., business licenses, payroll). These databases for the most part did not contain the data elements of interest to this study. For example, company business licenses name corporate officers, but not necessarily founders, and contain no biographical data. We next turned to University resources. The contracts and grants data, and some of that relating to technology licenses, was relatively straightforward to access and incorporate. The data about the people, however, was not. Alumni associations keep lists of their members, but employment data is self-reported, often outdated, incomplete and generally not reliable enough for this purpose. Similarly, academic departments rarely maintain systematic collections of this kind of information.

Data Collection: Founders

Information about company founders was derived from a systematic review of available corporate documents, including corporate profiles, prospects, California Department of Corporations filings, and 10-K forms. Data recorded includes the names and affiliations of both UC and non-UC founders. Information about private biotechnology companies is relatively difficult to find, and in general the names and UC affiliations, if any, of company founders are not readily available. Mark Edwards, Managing Director of the Recombinant Capital, generously provided access to his extensive collection of biotechnology company information, including corporate profiles, filings with the California Department of Corporations, and corporate prospects. Bill Otterson, Director of UCSD CONNECT graciously allowed CLP researchers to review the entire set of corporate profiles in the CONNECT archives, from which a number of UC linkages were identified.

To compile a systematic set of corporate information, a post card was sent to all "core" companies requesting that the Critical Linkages Project be placed on the corporate mailing list. The response rate for public companies was 100%, but the response rate for private companies was less than 25%, reflecting differences in legal reporting requirements and information dissemination strategies across the two groups. The information provided by public companies includes Security and Exchange Commission 10K reports, prospects, corporate profiles, and in some cases stock analysts' reports and press releases. For private companies, the information provided is generally limited to corporate profiles, occasional press releases and product catalogs. Advances in electronic databases and World Wide Web-based resources have allowed, since 1997, access to most information about public companies on-line.

Data Collection: Workforce

The absence of reliable alumni employment information from University or private sources convinced us that a direct survey of companies would be necessary. We created a simple survey instrument, consisting of a table with row headings, "Job Title," "Name," "UC Campus Attended," "Degree" and "Year." A copy can be found in Appendix A. Every "core" company was then contacted. First, a telephone call was made to the human resource director (or the equivalent) to introduce the Critical Linkages Project and to solicit preliminary information and a commitment to participate. All companies were given complete assurances that the data they provided would be held in strict confidence. Then, for companies agreeing to participate, a printed survey form was sent to the attention of the appropriate information source at the firm, generally the director of human resources. If completed survey forms were not returned within 3 weeks, follow-up telephone calls were made to encourage participation. For those who did not fill out the survey form, but did supply information by phone about UC alumni employment, letters were sent out requesting written confirmation of the information provided. Only information received from companies in written form was entered into the database.

Approximately 58% of companies surveyed (136 out of 228 companies) provided partial or complete responses to the survey. Eighteen firms declined to participate. Forty-four firms expressed initial interest but did not return survey forms. For the remaining 31 firms, the person responsible for human resource information could not be reached by phone.

Data Collection: Contract and Grants and Technology Transfer

In contrast to the foregoing categories of information, the University undertakes extensive tracking of contracts and grants and technology transfer activities, albeit with certain limitations. Information about contracts and grants funded by California biotechnology companies was provided by the UC Contracts and Grants Office. The set of research contracts funded by "core" firms was constructed by matching the CLP "core" list to the set of information on all research contracts in the UC Contracts and Grants database for the fiscal year 1994-95. The Phase I findings reflect grant activities reported in FY 94-95 only, although the Critical Linkages database will be updated to include data reflecting subsequent years. A major Project initiative currently underway seeks to reconcile the manner in which business entities, such as those included in this study, are tracked in the various UC databases with relevant data. For the initial work, a comprehensive manual review of all records involving private company sponsors of UC research in the Contracts and Grants database was undertaken to assure an exhaustive mapping between the two data sets.

Technology license information was provided by the UC Office of Technology Transfer for the licenses under their purview. Due to legal constraints on disclosure, information about individual licenses was not provided, but rather a count was made by comparing the CLP "core" list to the OTT list of licensees. Information about technology licenses from the UC-managed National Laboratories was collected via a phone survey of the technology transfer and industrial partnership offices at Lawrence Berkeley National Laboratory, Lawrence Livermore National Laboratory and Los Alamos National Laboratory. Data from the autonomous technology transfer offices at UC Berkeley, UC San Diego and UCLA remains to be collected.

Phase I Findings

The University of California is clearly a driver of the commercial biotechnology economy. The most striking results from the first phase of the Critical Linkages Project study include:

- 1 in 3 U.S. public biotechnology firms is located within 35 miles of a UC campus
- 85% of California biotechnology firms employ UC alumni with graduate degrees
- 1 in 4 California biotechnology firms were started by UC scientists, including three of the world's largest: Amgen, Genentech and Chiron, which alone accounted for 37% of the US biotechnology industry's 1996 total market capitalization ($19 of $52 bil.)

In the following section, we examine separately the results in light of the four distinct avenues through which knowledge generators contribute to the biotechnology economy; scientific leadership, workforce, scientific research and technology transfer.

Scientific Leadership

One of the most tangible ways that knowledge generators contribute to the emergence of new industries is through the founding of firms. Start-up activity by academic scientists has been the focus of intense interest by economists, including e.g., Zucker et al (1994, 1995), Audretsch and Stephan (1996) and others. Conceptually, there are two major categories of academic scientists who found biotech businesses. The first are those who have made a specific discovery that has commercial potential. Generally, the development of a marketable process or product requires extensive research efforts that are highly applied in focus, and thus not within the usual bounds of academic research. University scientists may seek to license their discovery to an ongoing concern or, if no such firms are interested, if license terms are not satisfactory, or if the scientist is strongly committed to entrepreneurship, start up their own venture to take on these developments. A second category consists of scientists who recognize a commercial opportunity to which they can apply their expertise, which does not involve their tangible intellectual property that would be subject to a license. For the purposes of this study, in part for reasons cited in the section below regarding technology licenses, we do not distinguish between these two subgroups of founders.

Initial Findings:

In the 1995 study, information available was sufficient to identify the founders of 98 California biotechnology firms out of the 228 "core" firms identified.[10] Of these firms, 58 were founded by UC faculty or alumni, 11 had both UC and non-UC co-founders, and 31 had no UC founders. This means that 68% of firms, in the subset of those with information available, had UC founders. More conservatively, it suggests that 25% of California biotechnology firms have been founded or co-founded by UC scientists.

The geographical location of founding activities by UC scientists has been quite consistent over time, as can be seen in Figure 1a. The majority of these firms have been located in the SF Bay Area, represented by the dark gray columns, followed by the San Diego region, shown in light grey, with somewhat fewer firms in the Los Angeles region, shown in black. These patterns closely mirror the overall patterns of start-up activity in the State, shown as the white columns in Figure 1b.

The first independent biotech firms were founded in the 1970's, the dawn of commercial biotechnology. Just 27 of the California biotech firms in existence in 1995 had been founded in 1980 or earlier, and of this group, 8 were founded by UC scientists. New biotech companies continue to be formed in California, a substantial subset by UC scientists. Of the 43 new companies we identified that had been founded in the five-year period 1991-95, 27 had UC founders. Overall, it appears that the proportion of companies with UC founders has not changed over time. This fact is somewhat surprising, because one might expect that, as the technology and skills of biotechnology become more broadly diffused over time, the advantage of academic scientists in founding firms might decline. In fact, the CLP data show that UC scientists remain an important source of new biotech companies.

A second means by which academic scientists contribute leadership to biotech firms is through service on scientific advisory boards (SAB). There were 108 UC faculty members serving on the SABs of 50 companies in the CLP study, including 7 who were designated as Chair or Co-chair, a position that may indicate particularly strong leadership contributions. An interesting question is the patterns of participation as SAB members and company founders; the current data is not sufficiently exhaustive to allow definitive conclusions to be drawn.[11]

Figure 1a. Regional Growth: Firms Founded by UC Scientists

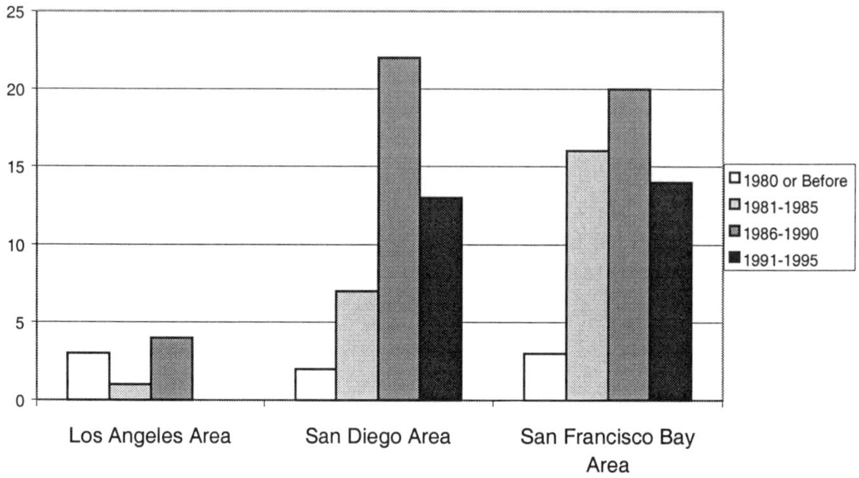

Figure 1b. Growth of Regional Biotechnology Clusters in California

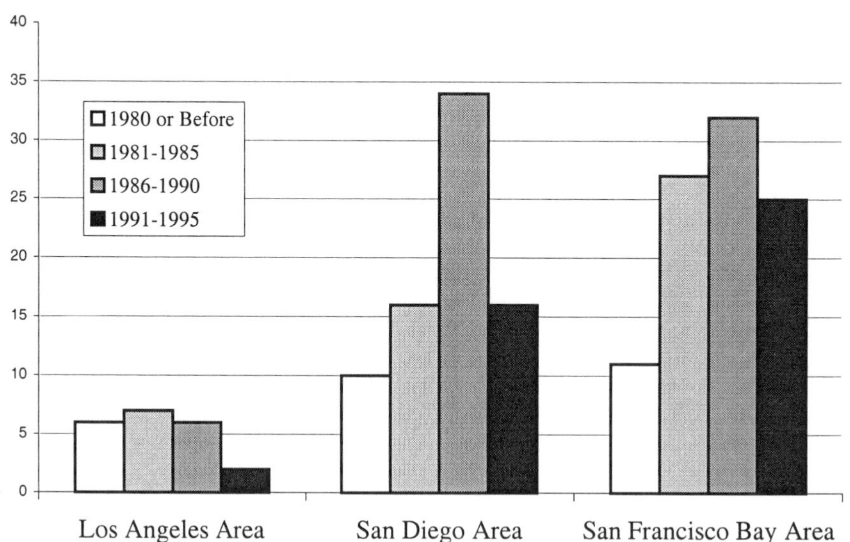

Workforce

The availability of a highly skilled workforce is another key element contributing to the success of the commercial biotechnology enterprise in California. As noted in the background section, there exists a strong correlation between the location and vitality of high tech business activity and the presence of college and university graduates in the labor pool. The employment of UC alumni, particularly in senior scientific positions, allows skills acquired at the University to be applied to the development of innovative new products and processes, providing the basis for high paying jobs for the alumni themselves, their staff, and affiliated businesses.

Initial Findings:

The importance of UC in educating the biotech workforce is evident from the responses received to the Critical Linkages employment survey. Responses were received from 134 firms, making up 58% of California biotechnology companies operating in 1995. Eighty-eight percent of those responding reported employing UC alumni, including 100% of respondents with more than 20 employees.

Of particular interest is the subset of employees who earned PhDs from UC. Figure 2a shows the proportion of UC PhD holders from each campus, of the 302 identified in company survey responses. It is notable that although the majority of these employees earned their doctorates at UC Berkeley, UC San Diego, UCLA or UC Davis, all nine UC campuses are represented. The job titles held by these alumni employees are shown in Figure 2b, which illustrates that UC graduate training contributes strongly to the senior executive ranks as well as the R&D activities of the surveyed firms.

Figure 2a. Campus source of UC PhDs employed by California Companies

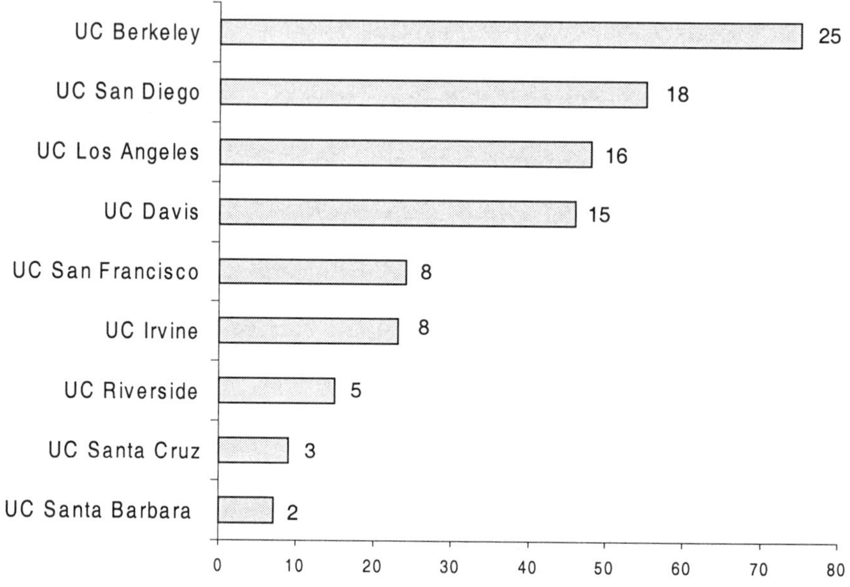

Figure 2b. Positions Held by UC PhDs

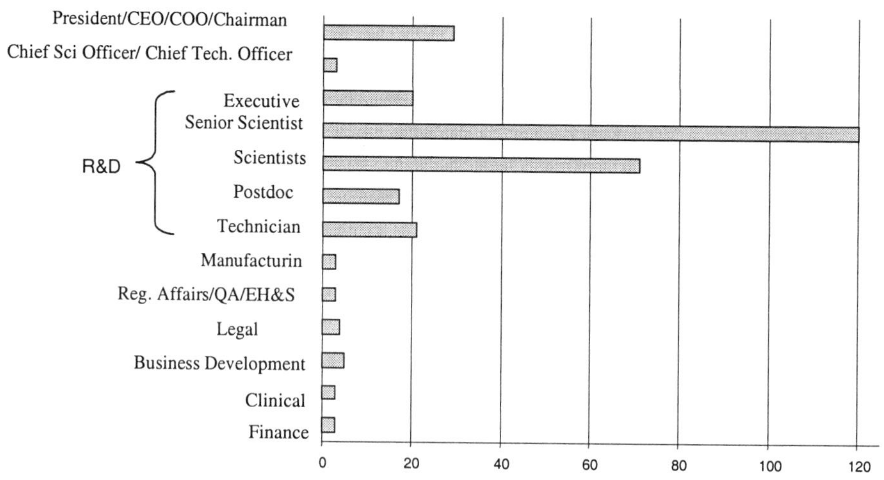

Research in Science and Engineering

In addition to scientific leadership and education, economic benefits accrue to research conducted by UC faculty and Laboratory scientists, which generates discoveries that add value to California biotechnology firms. Given the complexity of cutting edge technologies, individual firms often lack the resources to assemble and staff research facilities that can address all obstacles encountered along in a particular research path; access to the expertise and facilities of University researchers working on fundamental or applied research is essential to the success of such projects.[13] One indication of the importance of UC scientific innovation is the willingness of firms to support faculty and laboratory research projects.

Initial Findings:

Records made available by the National Laboratories and the UC Contracts and Grants office show that 27 California biotechnology firms funded 54 non-clinical faculty and laboratory research projects at UC in 1994-95.[14] Total funding provided for these 54 projects was $13.3 million. Table 1 shows the campus location of this sponsored research activity. An interesting line of future research would be to investigate the relationship between sponsored research and other linkages.

Table 1. Contracts and Grants sponsored by California Biotechnology

Campus	*Number of Projects*
UCSD	18
UCSF	13
UCI	9
UCLA	8
UCD	4
UCB	1
UCR	1

Technology Licenses

Technology transfer activities create economic value by facilitating the translation of University research into tangible commercial products and processes. Licensing of the innovations developed by UC researchers allows firms to establish property rights to the technology, so that they can recoup the investment in applied research, product development, manufacturing, marketing and distribution required to successfully bring the innovation to market. One indicator of the value of research conducted at UC to the biotechnology industry is the amount of money that firms pay the University to license UC technology. This indicator is subject to several important limitations. One major issue it that, because license revenues are typically linked to sales of a product, these revenues do not accrue until after the licensee has made significant investments in development and commercialization. This process may take years, which helps to explain why the vast majority of revenues are generated by a small number of "block buster" licenses. Another issue that makes reliance on licensing revenues a problematic measure of value is that licensees may make indirect contributions, such as providing other support for university research, that are not reflected in licensing revenue streams.

Initial Findings:

Licensing revenue for biotechnology inventions managed by the Office of Technology was $41m in FY 1994-95, representing 74% of the total licensing income. This pattern of licensing in which biotechnology and biomedical inventions dominate in terms of both number of inventions licensed and dollar amount of royalties collected, has been seen for all major US research universities. The University licenses technology to companies across the nation and the world, but for this study we looked at only at California biotechnology companies. In 1995, 82 of these companies held active licenses for UC technology. Seventy licensed the Cohen-Boyer patent, 32 held other UC licenses, and 20 companies held both types of licenses. The relative importance to a company of any given license is a question worthy of further research. Although not yet collected systematically, it was clear from the company documents reviewed for this study that many biotechnology businesses hold multiple licenses from a diverse array of academic institutions, research institutes and other businesses. Indeed, a major means of realizing value for many biotech companies is the licensing of technology to other firms. A systematic study of the role of licensing in business development, and the role of academic licenses in overall intellectual property portfolios could be a useful line of future research.

CONCLUSIONS

The major hypothesis underlying the Critical Linkages Project is that people, particularly those whom we have termed "knowledge generators," are the drivers on today's knowledge-based economy. To provide a concrete foundation for assessing the contributions of knowledge generators, we developed a framework in which we identified a comprehensive set of "core" biotechnology companies in California, then systematically undertook to identify and document the roles that University of California faculty and alumni have played, and continue to play at these firms. This approach strongly with that taken by most economic studies of the contribution of research and education to the economy, which generally have focused on a small set of indicators, especially publishing, patenting, and relative citation rates. While most such studies conclude that public investments in basic research and graduate training generate substantial economic benefits, they do not have a means to explain just what happens between the academic activities and the economic outcomes, leaving the process as a "black box." We have shown a light into this area, providing a unique, close-up view of that translation process. The Critical Linkages Project biotechnology case study clearly demonstrates that university faculty and alumni contribute to commercial biotechnology through a variety of channels, including starting up firms, serving as scientific advisors and key R&D personnel, conducting sponsored research and producing discoveries that are licensed by firms.

This study has examined a single sector of the economy, biotechnology, and looked solely at activities occurring in just one state, California, focusing on those relating to the public research university system. An important question for future research is whether the pattern of contributions by knowledge generators that we identified in this study can be generalized. Similar case studies will be needed in other high technology sectors before we can determine whether the framework we developed here provides useful insights across the economy more broadly. A study that examines one or more of the several other US or international biotechnology clusters would provide a means to assess whether the centrality of knowledge generators to the development of commercial biotechnology that we find in California generalizes. The data collection effort undertaken for this study was substantial; if this framework is to be useful in a larger context, efforts will have to be made to find effective ways to streamline and simplify reporting of relevant information. Finally, we have just scratched the surface in asking questions of the dataset for this case study. Many interesting questions remain to be investigated, including questions about relative performance of faculty-founded companies, variation in patterns of relationships (licensing, contract research, alumni employment) in faculty-founded versus other companies, and the impact on faculty members academic career paths of relationships with companies.

ACKNOWLEDGMENTS:

The following people and organizations have provided valuable contributions to the Critical Linkages Project: Prof. Lynne Zucker, Institute for Social Science Research, UCLA; Prof. Michael Darby, Anderson School of Management, UCLA, UC Systemwide Executive Committee members, Prof. Suresh Subramani, Dept. of Biology, UCSD; Dr. Greg Lennon, Lawrence Livermore National Laboratory; Dr. Paul Jackson, Los Alamos National Laboratory; Barbara Yoder, Lynn Judnich, Suzanne Quick and Joe Acanfora, UC Office of Technology Transfer. Alameda; Delia Talamantez, UCSD Conflict of Interest Office; David Gilbert, Vivianna Wolinksi, and Jeff Weiner, Lawrence Berkeley National Laboratory; John Mott, Los Alamos National Laboratory; Mark Edwards and Satomi Degami, Recombinant Capital, San Francisco; Bill Otterson and Dr. Abi Barrow, UCSD CONNECT; Doug Windsor and Jean Cunicelli, Ernst and Young, Palo Alto; Jeannine Niacaris, Bay Area Personnel People in the Life Sciences; William Hoskins, UC Berkeley Office of Technology Licensing; John Tucker, UCSF; and the alumni associations and related University offices at all nine campuses.

I am indebted to Fred Gault, Paula Stephan, David Mowery and the participants at the Statistics Canada/PRIME Advanced Research Workshop on the Economic and Social Dynamics of Biotechnology (Ottawa, February 24, 2000), and the University of California IUCPR Economic Indicators for Academic Science Workshop (Berkeley, March 2000) for their comments and suggestions. This study was funded by the University of California Biotechnology Research and Education Program, and subsequently the UC BioSTAR Project. Susanne Huttner, director of both programs, has been instrumental in providing support and direction for the project.

ANNEX

UC Alumni Employment Survey

Company Name: ______________________

Name: ______________________
Title: ______________________
Date: ______________________

	Job Title	Name	UC campus attended	Department	Degree	Year
1						
2						
3						
4						
5						
6						
7						
8						
9						
10						
11						
12						
13						
14						
15						
16						

Number of UC Alumni: ____________ of total number of Employees: ____________

*** Please fax completed form to: Critical Linkages Project, UC Biotechnology Program, Fax # 510-643-2574*

PART IV

IMPACTS

Chapter 9

INTERNATIONALLY COMPARABLE INDICATORS ON BIOCHNOLOGY:

A Stocktaking and Proposal for Work

Bill Pattinson, Brigitte Van Beuzekom and Andrew Wyckoff
Directorate for Science, Technology and Industry, OECD

INTRODUCTION

Although work on measuring biotechnology is just beginning within a select number of countries, many policy makers and analysts are calling already for an international effort to co-ordinate this work so that the ensuing statistics and indicators maintain some level of international comparability. Given the early stage of this technology and all the other competing priorities for statistical work (e.g. on mature but important areas such as services), this demand may seem misplaced but it is reflective of 3 events that are quickly evolving on the world stage: 1) issues associated with the trade of goods and services that involve biotechnology; 2) the belief that biotechnology will be a source of economic growth and a factor that may determine the relative competitiveness of countries, and 3) the recognition that this technology is a "general purpose technology" where the impact will be felt widely across industries and countries, regardless of its origin. These issues imply that the nature of measurement needs to shift from private sector sources mainly concerned with market growth to public sources that can place biotechnology into a broader framework including international trade, innovation and productivity which entails the need to directly compare measures of biotechnology to other publicly generated data.

For these reasons and because it has been successful at co-coordinating the measurement of other conceptually and mechanically difficult areas such as research and development, innovation, and growth of the information society, the Organisation for Economic Co-operation and Development (OECD) was asked to explore the feasibility of compiling statistics and generating indicators of biotechnology for its Member countries.

With the support of the Government of Canada, this paper presents an initial stocktaking of the current state of methodological and statistical work on biotechnology by identifying existing national statistics, compiling a set of definitions currently in use and surveying user needs as identified in a number of countries. While somewhat descriptive, this stocktaking represents a foundation upon which future work to achieve international standards and measures can be built.

EXISTING NATIONAL STATISTICS

To help assess the availability of data in OECD Member countries, a questionnaire was sent to assess the availability of statistics and indicators, the methodology used in collecting this data (particularly the definition of biotechnology used) and quantitative analyses undertaken on biotechnology. While this questionnaire was sent to government officials who attend OECD working parties related to these issues, they were asked to provide responses to these inquiries regardless of whether or not this work was performed by a government agency. A summary of replies received from 21 member countries is available in Pattinson et al. (2000).

The most common form of official data collection came in respect of R&D data collected as a part of their national R&D surveys. Five countries – Australia, Canada, Denmark, Japan and the Netherlands – collect data in this way. The data items available, however, are very limited as they generally only relate to R&D expenditures and the human resource input to that R&D. Only two countries have specific biotechnology firm surveys – Canada and France. A third, New Zealand, is proposing to undertake a similar survey. These surveys are quite detailed and measure many data items about the structure and performance of the firms using biotechnology processes, producing biotechnology goods and services or undertaking biotechnology research.

A start-off practice, undertaken by some national statistical offices, has been to commission consulting firms to undertake national audits of the biotechnology industry. This has notably been the case for Australia (Ernst & Young), Canada (Ernst & Young) and France (Arthur Andersen). In the case of the latter two, the national statistical offices subsequently refined the commissioned surveys, narrowing the sampled population as well as the statistical questions.

Leading consulting firms such as Ernst & Young and Arthur Andersen have, for many years, been producing reports that track developments in specific countries. These publications generally have a fairly broad scope. Ernst & Young, in particular, has produced reports for Europe, the U.S., Germany and Belgium (forthcoming – in conjunction with the

Belgian Biotech Association). The US report is in its 14th year of publication and the European in its 7th year. These consultant firms also compile data on indicators such as venture capital in some countries.

Some of the more prominent pharmaceutical and agricultural corporations, (Monsanto, Wellcome, Dupont, etc.) also provide a range of information and indicators about biotechnology on their Internet web sites. A non-exhaustive list of relevant web sites and publications is given in Pattinson et al. (2000).

Most countries have national biotechnology associations. These associations provide a forum, to bring together all the players involved in the biotechnology community. These are important sources of information because they maintain registers or directories of public and private enterprises. Biotechnology associations often work co-operatively with public sector organisations to provide statistical information. In Australia, for example, the online 1999 Directory was produced jointly by the Australian Biotechnology Association Ltd and the Australian Government Department of Industry, Science and Resources. In Japan, the Japanese Bio-industry Association conducts its own survey measuring a range of statistical indicators including the value of biotechnologically produced goods.

The European Commission produced an inventory of public biotechnology programmes in all countries participating in the EU biotechnology 4th Framework programme 1994-1998 (17 countries). These reports present inventories of "specific aspects of each programme and other policy instruments by which biotechnology has been stimulated and additional information with respect to the national conditions, policies and impacts, etc".

Hungary and Finland have also conducted evaluations or "audits" of the biotechnology sectors that have enabled the compilation of some statistical indicators in those countries.

In addition to this directly collected data, there is other information that becomes available as a result of administrative records, most notably being patents data (from national and international Patents offices) and international trade data, generally coming available from national Customs offices.

THE STATISTICAL DEFINITION OF BIOTECHNOLOGY

The definition of biotechnology used by responding countries varied considerably - as noted above - but by and large consisted of two types – one based on a set of words which describe the processes referred to, called a "single" definition and a list-based definition. Each has its uses and both are required for different statistical purposes. Examples of both are given in Pattinson et al. (2000).

A broad definition is useful for measuring biotechnology using existing surveys on R&D or industrial activity generally. But in addition to a single broad definition, sub-classifications need to be developed to enhance the value of any statistical outputs that would be derived. Such classifications may be part of existing classifications used in existing statistical practices, such as standard industrial classifications or socio-economic classifications, or may need to be developed specifically to enhance the understanding of biotechnology. This issue needs to be further taken up in the development of an overall statistical framework that incorporates, inter alia, definitions, data items and classifications.

The list approach is used in diffusion surveys such as those conducted by Canada (as discussed by Antoine Rose above) and France. In general these list use a detailed set of biotechnological process, but they need to be sufficiently flexible, generic and broad, as growth in biotechnologies is so rapid that any fixed list would be quickly rendered obsolete.

User Requirements

A prerequisite for the development of internationally comparable indicators is the preparation of a set of user requirements, specifying the policy issues which needed to be addressed and the indicators that might be useful for that assessment. An initial list could include the following indicators, each of which are discussed briefly below:

- R&D, innovation and linkages between sectors in national innovation systems.
- Human resources devoted to biotechnology.
- Stock of biotechnology human resources.
- Patents and citations.
- Venture capital.
- Biotechnology diffusion.
- Production and trade in biotechnology products.
- Structural information about biotechnology firms
- Business start-ups.

R&D, Innovation and Linkages Between Sectors in National Innovation Systems

This is a key area for biotechnology statistics. It is important to know about the amount of R&D that is being carried out, who is conducting it, who is funding the R&D, the purposes of the R&D, and similar indicators. Further it is important to understand how this R&D leads to innovation, in the form of new products and processes.

As with most R&D statistics, data is required about the expenditure on such R&D (and innovation) and the human resource inputs to it. It is also important to understand the linkages between the R&D, and the R&D firms, and other aspects on the innovation systems occurring within Member countries. In particular, the linkages between the different institutional sectors need to be identified and measured, particularly if these links extend beyond a country's borders.

One mechanism for integrating the measurement of into R&D and innovation surveys is to include it in the Frascati (OECD 1993) and Oslo (OECD 1997) Manuals, that respectively set measurement guidelines for OECD Member countries in the areas of R&D and innovation.

Human Resources Devoted to Biotechnology

Indicators on human resources devoted to biotechnology are among the most difficult to compile. While there can be indicators relating to human resources inputs to R&D and innovation coming from those type of collections, this omits any requirement for human resources involved in the application of biotechnology within industrial and other processes. These are potentially available from technology use surveys and hence it may be necessary to combine data from a range of collections if one is to develop an accurate measure of total human resources devoted to biotechnology. Of course taking data from different sources leads to potential problems with overlap and gaps between them.

Stocks of Biotechnology Human Resources

Human resources qualified in biotechnology, but not necessarily in a scientific or biotechnology occupation, require data from different sources. Some countries have Population Censuses that are able to provide such data; these of course tend to be very infrequent data points. Other countries have population (or household) surveys that can also be used for such purposes but often the data from these is unavailable at that level of detail. In other countries there is data available from administrative records. The OECD and Eurostat have jointly developed a statistical manual (the Canberra Manual, OECD 1995) which aims to address this question for highly qualified people in general; the revision of this manual may provide an opportunity to include a demand for such data in the revision of that Manual.

PATENTS AND CITATIONS

There seems to be a demand for patents and citations data in respect of biotechnology processes. There is a significant body of data that is available from national and international Patent Offices and these ought to be used to provide indicators. The OECD Patent Manual (OECD 1994) provides guidelines for compiling and interpreting the data.

Venture Capital

Venture capital is an important determinant of the biotechnology industry and business start-ups in biotechnology. There is very little official information available; however, there is a considerable body of statistics which are put together from private sources.

Biotechnology Diffusion

This is a most important aspect of a framework for biotechnology statistics; there are examples of such surveys from Canada and France and shortly from New Zealand. Within such surveys it is possible to address many of the user requirements for statistics on biotechnology. However they are by nature one-off and quite expensive to conduct. Thus while such surveys are an integral part of an overall framework, they need to be supplemented by other options.

Production and Trade in Biotechnology Products

The measurement of trade and local production in biotechnology products is a more difficult statistical issue. There are no readily-available lists of firms that undertake the production of biotechnology produced goods; clearly there are some industries that can be targeted for the measurement of relevant products and this might be an acceptable approach. International trade, however, is a much more difficult issue. International trade in products is generally measured via administrative records and these are unlikely to be suitable for use in identifying trade in biotechnologically produced goods, at least in the short to medium term.

Business Start-Ups

Business start-ups is an indicator that is frequently used, particularly to measure the growth of new emerging "industries". It is however one which is very difficult to measure, as most countries have significant difficulty measuring business start-ups in general. Nevertheless it is an area which is of great importance from a political viewpoint and needs to be incorporated into an overall statistical framework.

Structural Information About Biotechnology Firms

Much debate goes on about the size of the biotechnology industry; but is there such an industry? Generally, in standard industrial classifications businesses are classified to an industry on the basis of their outputs and their production processes. So, logically, there is no industry classification which specifically identifies biotechnology, as this a process. The outputs of a business are not split between whether they are biotechnologically produced, or otherwise.

Thus while it may not be possible to define a biotechnology industry in the normal way, there will be a demand for information about the overall size of the sector which is involved in biotechnology and this will need to be accommodated in the overall framework.

CONCLUSION

Work to date has shown that there is very little data on biotechnology available from national statistical sources, with the exception of Canada and to a lesser extent, France. While this severely limits the nature of analysis that can be undertaken, it does present a good opportunity for countries contemplating the collection of biotechnology statistics to benefit from the experience of these leading countries, which in turn should ensure a greater degree of international comparability. By sharing experiences on what works and does not, a solid set of data can be constructed internationally that can be used to analyse the various issues that will be associated with this technology. Even though statistical work is at an embryonic stage in individual countries, it is important that an international dimension be initiated now as well. This will avoid the situation where we now find ourselves in the area of indicators for the information society where 20 years after the invention of the PC we still do not internationally comparable indicators of ownership and use of PCs in households. While not a resource-free proposition, if initiated soon, this work will be far less costly than if launched 20 years hence.

Chapter 10

THE CONTESTED INTERNATIONAL REGIME FOR BIOTECHNOLOGY:
Changing Politics and Global Complexity

G. Bruce Doern
CRUISE, Carleton University

INTRODUCTION

The purpose of this chapter is to explore the evolution and changing nature of the international regime for biotechnology and indeed the extent to which an international regime actually exists for biotechnology. It also explores the core international and domestic politics shaping the regime. The notion of a regime must be taken as a question because the nature of international governance is that there is no world government. For our purposes a regime is an interacting set of organizations, statutes, agreements, ideas, interests, and processes engaged in policy development, rule making and implementation in a policy field. In other words, the first test of there being a regime is that there is some inner core of such features and characteristics that warrant such a designation for analytical purposes.[1]

For biotechnology, it is also important to appreciate that the regime analysis has to explore separately aspects of the notional regime that are related to biotechnology in food (genetically modified or GM food) on the one hand; and biotechnology in health and therapeutic products on the other (DaSilva and Ratledge, 1992). Moreover, the international aspects of the regime inherently involve interactions where states are both regulating biotechnology and promoting and supporting its advancement. The regulatory aspects typically involve several interlocking systems of rules and norms. The promotional aspects are partly build into the same regulatory elements but also extend to varied and active state R&D support such as funding for the Human Genome Project but also through other support for medical research in general.

Inevitably in such a brief chapter, we can only sketch out a few basic features and key issues.[2] There is more of a focus on the regulatory aspects of the international regime for biotechnology than on the state support aspects but both must be kept in mind. The analysis proceeds in five sections. The first four sections map out key elements of the notional international regime for biotechnology and some of related politics which forged them: the World Trade Organization (WTO), its dispute settlement processes and approaches to science-based regulation, and the Trade-Related Agreement on Intellectual Property (TRIPs); food and pharmaceutical regulatory approval processes; the biosafety protocol and the importance of environmental criteria and the precautionary principle; and aspects of state support for biotechnology and informal and formal networks among scientists.

In the first two regime elements, biotechnology can be seen as simply a subset of its rules and norms and indeed key interests and countries have tried to ensure that biotechnology is seen as just another product or technology rather than something that requires special treatment or its own regime (Mironesco, 1998). In the third and fourth elements examined, the biosafety protocol and state support activities, there is already the notion that biotechnology deserves its own designated "regime" to highlight its differences as an enabling technology and set of products and processes (Ashford, 1996).

The final section then briefly examines the core overall politics of the regime-building process in the last decade, especially between, the U.S, Europe, and developing countries. This section includes recent developments where the global and national political debate is increasingly dividing along the food versus health fault lines of the politics of biotechnology. Conclusions then follow.

It must be stressed that the analysis of international regimes for biotechnology is filled with analytical "boundary problems". First, as the four elements suggest, biotechnology is in a real sense not governed by international or national governance structures that have been designed only for biotechnology Instead, biotechnology is an "add-on" to existing governing structures. This is true even though there is increased recognition that we may well be embarking on a "biotechnology century" (Rifkin, 1998; Appleyard, 1999; Grace 1997). Second, as already noted, the major fault lines of the politics of biotechnology are dividing increasingly between biotechnology in food and biotechnology in health, with the level and tenacity of public opposition and criticism much higher about the former than the latter (Baker, 2000). Third, biotechnology has gradually emerged on the national and international policy and economic agenda in the 1980s and 1990s in varied ways: in the development of national biotechnology strategies in some countries; in an evolving biotechnology regulatory system responding to industry's development of new products and processes and out of concern for longer and more secure trade and intellectual property rights; and in periodic controversies about products and scientific developments such the rbST, and the cloning of Dolly the sheep (MacDonald, 2000; Wilmut, Campbell and Tudge, 1999). All of these aspects of the politics of biotechnology have occurred at differing rates, and with varied salience in media coverage and political attention, in key countries and jurisdictions, such as the United States, the European Union, and developing countries (Abbott, 1989; Achara, 1992; Bhat, 1996; Shiva, 1997).

THE WTO: DISPUTE SETTLEMENT, SCIENCE AND TRIPS

The emerging notional international regime for biotechnology is a part of the larger WTO system including the WTO-centred Trade-Related Intellectual Property (TRIPS) Agreement. We cannot possibly describe or deal with the whole WTO apparatus (Lal Das, 1999; Trebilcock and Howse, 1995). Biotechnology is simply one unnamed embedded product or process that can be traded and hence might be caught up in the larger WTO trade rules and processes. But three key WTO aspects are central to the notional biotechnology regime: WTO dispute settlement; its science-based rules; and TRIPS.

First, the dispute settlement system under WTO has more teeth that the earlier GATT processes and is intended to work more expeditiously (Petersman and Marceau, 1997; Trebilcock and Howse, 1995). Dispute resolutions have a strict time limit established for the conclusion of the process and a single member country is prevented from blocking the adoption of reports of trade-dispute panels, or, on appeal, of appellate bodies. The WTO is also more pro-active because it has a trade policy review mechanism in which it has an independent investigative authority to initiate rotating country-by-country reviews of international and domestic policies that might adversely impact on trade relationships among countries. At the insistence of the U.S. and the EU, these core mechanisms were intended to entrench liberalized trade and minimize obstacles to trade, including trade in biotechnology products.

A second WTO feature centres on ideas about science in international trade (Browne, 2000; Lal Das, 1999; Trebilcock and Howse, 1995). Core WTO and other trade provisions (e.g. NAFTA) are anchored on the importance of objective and transparent science (in short, credible evidence) to underpin international regulation and standard-setting in health, safety and environmental matters. Science-based regulation is seen as crucial to ensuring that health, safety and environmental rules are not trade-distorting and do not become the new guise for protectionism (Doern and Reed, 2000; Vogel, 1995). There has always been a tension in trade agreement negotiations between environmental criteria where the precautionary principle is advocated (see below) and trade-economic criteria where a higher threshold of science or scientific evidence is advocated or agreed to in negotiations.

Trade-related ideas about science have emerged in debates about eco-labelling and in the issue of whether campaigns led by environmentalists for, in effect, competitive informal standard-setting (such as a consumer boycott) undermine official science-based standards sanctioned through trade agreements (Browne, 2000). Such informal efforts are seen by some as a threat to democratic, formalized international science-based regulation and harmonization whereas, clearly many other groups and environmentalist coalitions see it as a necessary democratic counterweight to the institutionalized or official trade system.

The third aspect of the WTO, though linked to other international norms and institutions as well is TRIPs. Under the pre-Uruguay GATT framework, the international regime for intellectual property (IP) fell well short of a harmonized regime (Doremus, 1996; Drahos, 1996, 1997; Sell, 1998; Doern, 1999); Indeed, IP issues were largely outside the GATT purview such as regarding Most Favoured Nation provisions and aspects of National Treatment principles as well. Moreover, international IP organizations such as the World Intellectual Property Organization (WIPO) did not contain a formal court-like process for dispute resolution. It regularly reported on disputes but had no GATT-like panel process for dispute resolution. (Doern, 1999; Trebilcock and Howse, 1995). The Uruguay Final Round Act included for the first time a comprehensive agreement on TRIPs that seeks to balance the conflicting values inherent in IP and between developed and developing countries (Bhat, 1996). It greatly strengthened the role of the WTO but it also established a new body, the Council on TRIPS and mechanisms to help developing countries get ready for the new stricter regime.

During the Uruguay Round, the issue of mandates in IP between the WIPO and the proposed World Trade Organization generated considerable dispute (Doern, 1999). The developing countries preferred the WIPO as the lead institution because, it had facilitated diverse IP policies and institutions in developing countries. The United States and Europe, but especially the former, wanted a stronger WTO mandate because it wanted better dispute settlement and enforcement of IP rights, especially regarding key developing countries whose IP regimes were either weak in law or weak in their implementation. As a result, the Uruguay Final Round Act includes for the first time not only the WTO itself but also a comprehensive agreement on TRIPS that, in principle, seeks to balance the conflicting values inherent in IP and between developed and developing countries.

But central to WTO-TRIPS is also the achievement in the Uruguay Round of a harmonized patent of twenty years duration (from filing date). The length of patent protection is an issue replete with both economic and political calculation and pressure. A key question is "why twenty years?" and another is "why one period for all industrial sectors or kinds of invention?".

Patents are intended to produce a temporary monopoly to reward intellectual effort and ingenuity. But simple economic logic suggests that these periods of protection ought to vary greatly by field or sector depending on varying cost structures, investments, and payback periods. This also suggests that countries would have different views about what kinds of protection across sectors would make the most sense given their national state of development and strategies for development. Thus the underlying economics of patent protection suggests the suitability of many periods of protection and that these could also change over time.

However, the political and institutional logic is somewhat different. First, for key players the basic logic is simply the longer the protection period the better. This view is driven by firms such as those in the national and global pharmaceutical and biotechnology industries who sought out and achieved maximum effective protection. Their desire for maximum periods is driven by factors such as high upfront costs in R&D and in obtaining ever lengthening drug approval processes by other government regulators in several countries. In the 1980s they saw their effective protection being reduced and sought change in national laws and trade regimes. The "longer is better" logic was also the driving force behind the United States, and later the European Union, in successive trade negotiations. U.S. power was crucial in this regard in that the Americans saw IP as increasingly crucial for American economic development in "new economy" industries such as biotechnology and related it both to developing countries with weaker regimes on patents but also to fellow-developed countries such as Canada which it pressured to change its patent laws as well (Trebilcock and Howse, 1995). In Canada, this pressure focussed on Canada's preferences given to generic drug manufacturers and was brought to bear both before and during successive FTA, NAFTA, and GATT negotiations (Doern and Sharaput, 2000).

In the last decade, in particular, there were few if any effective counterpressures from those interests/countries that might have made the counter arguments. Developing countries mounted some counter pressure but were eventually worn down by more powerful forces. Consumers in some overall sense had a vested interest in less monopolistic practices but, at both national and certainly at international levels, they were a weak, diffused, and virtually voiceless interest. Perhaps the only exception to this was in the health sector where health ministries were often a surrogate representative of consumer or patient interests.

And last but hardly least, the logic of longer protection periods and for one long period also came from the political and administrative logic of trade negotiations and implementation. It was simply easier to agree on one such longish period because it would be easier to implement. Moreover, it would preserve the notion that IP was indeed an area of real framework law which applied across the economies of member states and did not constitute a form of sector-specific "industrial" policy which it would be if many sectoral-based periods of protection were possible.

As mentioned, the TRIPS regime dealt with IP as a whole and hence biotechnology is nominally only a part of the backdrop. But pharmaceutical industries and related biotechnology firms were the core lobby in the U.S. pushing for the new regime and they have largely succeeded (Sell, 1998).

However, a further area of patent policy was more overtly biotechnology- specific. This area centres on the patenting of higher life forms (as opposed to microbial life forms which are patentable in many countries) and the broader economic and ethical issues inherent in regulating biotechnology. U.S. law allows such patenting whereas EU and Canadian law does not or is considering the need for special procedures. Debate here turns on what constitutes an invention and the degree to which the manufacture or the composition of matter was under the control of the inventor as opposed to the laws of nature.

PHARMACEUTICAL AND FOOD REGULATION

The second key element of the notional international regime on biotechnology is centred on the processes and institutions which regulate and approve new pharmaceutical and food products. At its core this element proceeds from a series of *national* health and safety regulatory bodies rather than from a fully developed international approval process as such. However, WTO and trade-related dispute settlement processes are ever more influential in this realm as well.

The WTO Agreement on the Application of Sanitary and Phytosanitary Measures (the SPS agreement) places disciplines on national processes of regulation and enforcement regarding the protection of human, animal or plant life or health from risks arising from animal or plant pests or diseases, food additives or contaminants (Lal Das, 1999; Browne, 2000). The intent here is to prevent the use of SPS measures from becoming disguised restrictions on trade while safeguarding each country's right to protect health and safety. The key principles of SPS measures are: avoiding unnecessary obstacles to trade; basing measures on scientific principles, scientific evidence and risk assessment; harmonization; and equivalence and transparency.

Another international regime element is the CODEX Alimentarius Commission. In 1989, it decided to evaluate applications of biotechnology under the existing CODEX system, which meant that no separate committee was established. Thus, in the both SPS and CODEX aspects as a whole, the intent was that biotechnology was not singled out. It was just another product which like others had to be looked at using sound science-based regulation.

But the real politics of this aspect of regime construction is found in the nature of national regulators or rather sets of national sectoral regulators. Canada is a useful example here. When the biotechnology industry emerged and assumed prominence, a biotechnology strategy was devised but the approach taken to regulating biotechnology products was to decentralize it within existing areas of regulation such as those which handled food, health, fish products, and environment (Doern and Sheehy, 1999). In Canada, legislative responsibility for biotechnology is divided primarily among four departments: Environment Canada, Agriculture and Agri-Food Canada, Health Canada and the Department of Fisheries and Oceans. In addition, Industry Canada partially coordinates policy development activities related to a biotechnology strategy across government departments and has been a strong force within the government in seeking to establish and strengthen an internationally competitive biotechnology industry.

The case handling and approval process of the Canadian federal biotechnology regulatory system is complex. The biotech producer goes first to the relevant sectoral regulator in the agricultural, health or fisheries departments, with the environment department and its relevant legislation held in reserve as the safety-net regulator. The biotechnology producer's entree to the sectoral regulator is through the provisions of the relevant sectoral statute covering the product.

In national terms, much of this approach made sense in that there were sound reasons as to why a government might not want to create a stand-alone "biotechnology regulator". Both key industrial interests and the many regulatory institutions (except environment departments) opposed any designated biotechnology regulator. But problems emerged when key cases such as the rbST case came on the regulatory agenda. In both Canada and the EU, these cases became controversial enough to raise larger political concerns not only about these products but also about biotechnology as a larger enabling technology (MacDonald, 2000; Ashford, 1996).

International regime development must also be linked to industry efforts to benchmark the performance of national regulators in areas such as pharmaceutical regulation. In the 1990s in particular, the U.S. and European-led global pharmaceutical and biotechnology industry has exerted pressure on national regulators to enhance the efficiency of the regulatory systems (Vogel, 1998; Doern and Reed, 2000). Some of this pressure has crystallized recently around an effort by the independent, UK-located, primarily industry-funded,

Centre for Medicines Research International (the CMRI) to develop agreed benchmarks or "league tables" for the approval times of various national regulators (Centre for Medicines Research International, 1997). Several countries are cooperating in this effort through information exchange. In this regard the CMRI process has led to the need to identify what the key milestones are in the drug review process. Without such agreement, one could not compare data on approval times. The 1997 CMRI report concluded that "over the 1990s, improvements have been seen in overall review times for a number of authorities including Germany, Spain, Australia, Canada and the USA" (Centre for Medicines Research International 1997, p. 13). But the report also shows that only six milestones were identified as being applicable to the review processes of all 9 countries being studied and "only one milestone is routinely recorded by member authorities" (Centre for Medicines Research International 1997, p. 12).

THE BIOSAFETY PROTOCOL

A third element of the international regime for biotechnology has been centred on the negotiation of the biosafety protocol to the UN Convention on Biological Diversity.

The crucial event and process here starts with the forging of the Convention at and after the 1992 Rio Earth Summit (Purdue, 1995). The process became an intense North-South economic and political issue. In the early 1980s, developing countries had been able through the Food and Agriculture Organization (FAO) to obtain an undertaking which said that all seeds are a common heritage. These included "inbred elite lines used for breeding by seed companies" (Purdue, 1995, p. 101). Developing countries by the late 1980s had countered this with an agreed FAO interpretation that intellectual property rights were not incompatible with the earlier undertaking. The developing countries shifted their positions somewhat as Rio approached, seeing seeds and plants as natural resources over which nations had sovereign authority. The resulting Convention speaks of the need for adequate and effective IP protection and hence in the view of many critics ensures that the WTO-TRIPs agreement will prevail over the biodiversity agreement.

In November 1995, parties to the Convention began work on a draft protocol on biosafety, which specifically focussed on the transboundary movement of any living modified organism (LMO) resulting from modern biotechnology that might have adverse effects on the conservation and sustainable use of biological diversity. The negotiating mandate pointedly excluded issues related to the safety of genetically modified food. Its focus was on the transboundary movement of *living* modified organisms, i.e. those capable of reproduction and not the non-living products derived from them.

Several meetings and negotiations were held, culminating in the adoption of a consensus text in Montreal on January 29, 2000. Fifty ratifications are needed to bring the protocol into force. In the final phases of negotiations, five negotiating groups emerged: the "like minded group" composed of most developing countries; the "Miami group" composed of six major agricultural exporters (Canada, the USA, Australia, Argentina, Chile and Uruguay); the European Union; the "compromise group" (Switzerland, Norway, Japan, Korea, Mexico, Singapore, and New Zealand); and Eastern Europe, Russia, and countries of the former Soviet Union.

The key features of the protocol are as follows (UNEP, 2000; Environment Canada, 2000; U.S. Embassy, 2000). First, as mentioned, the scope of the protocol is limited in that it applies to the transboundary movement, transit, handling and use of all living modified organisms that may have adverse effects on the conservation and sustainable use of biological diversity, taking also into account risks to human health. The human health reference is interpreted as health effects from environmental and occupational exposure and those resulting from an adverse impact on biodiversity. As mentioned food safety is not addressed. Pharmaceuticals for humans addressed by other relevant international agreements or organisations are exempted and only some of the provisions of the protocol apply to LMOs in transit or in "contained use".

A second feature is the application of Advance Informed Agreement (AIA) procedures to imports of LMOs. These apply to the <u>first transboundary movement</u> of LMOs destined for <u>deliberate release into the environment</u>, e.g. seed, saplings, fish and micro-organisms for bio-remediation purposes. Decisions on import are to be taken on the basis of risk assessments and pursuant to specified procedures and time frames. A signature country's failure to acknowledge receipt of a notification or to communicate its decision on import within the prescribed time frames does not imply consent to the import.

A third feature of the protocol deals with advance information on domestic approvals of LMOs destined for food, feed and processing "bulk commodities". This advance information would allow countries to determine and advise what regulatory requirements, if any, would apply to the first import of such LMOs well in advance of these LMOs entering into international trade. Another provision ensures that the documentation for all such shipments must clearly indicate that the shipment "may contain" LMOs which are not intended for intentional introduction into the environment and a contact point for further information.

A fourth element dealt with trade with non-parties. It provides that trade in LMOs between parties and non-parties must be "consistent with the objective of the protocol". This provision is of importance to several parties regarding trade with the U.S. This arises from the fact that the U.S. will not be a non-party for some years as it cannot ratify the protocol until it ratifies the Convention.

A fifth aspect of the protocol deals with socio-economic considerations in decisions on imports. Parties may take into account, consistent with their international obligations, socio-economic considerations arising from the impact of living modified organisms on the conservation and sustainable use of biological diversity, especially with regard to the value of biodiversity to indigenous and local communities. This means that socio-economic considerations can be taken into account in risk management but not risk assessment.

In addition, and crucially, the preamble to the protocol recognizes that trade and environment agreements should be mutually supportive with a view to achieving sustainable development. It emphasizes that the protocol should not be interpreted as implying a change in the rights and obligations of a signatory under any existing international agreements. The protocol incorporates the dispute settlement provisions of the Biodiversity Convention but parties preserve their right to have trade disputes arising out of a country's implementation of the protocol's provisions resolved in the WTO.

The protocol also contains references to the precautionary approach. Principle 15 of the Rio Declaration approved at the 1992 UN Conference on Environment and Development is referred to in the preamble and in the objectives article. The precautionary approach is also "operationalized" in an effort to ensure that the application of the precautionary approach by a party of import is understood to be linked to a science-based process for taking import decisions.

The precautionary approach or principle is a central paradigm which has currency and importance in debates about biotechnology for a number of reasons (Stirling, 1999; European Commission, 1998). In general it implies that action be taken by policy makers and regulators to prevent environmental damage even if there is uncertainty regarding its possible cause and possible extent (Connelly and Smith, 1999). According to this view "the environment should not be left to show harm before protective action is taken; scientific uncertainty should not be used as a justification to delay measures which protect the environment" (Jordan and O'Riordon, 1995, p. 59). One implication of this view is that one does not need complete scientific proof to practice the principle. As Jordon and O'Riordon argue, the setting for the use of the precautionary principle might be more propitious in recent years because of a broad-scale critique about science in environmental policy and regulation. They argue that the "science of assimilative capacity, predictive modeling and compensatory investment to offset the loss of ecological resilience is being challenged" (Jordan and O'Riordon, 1995, p. 61). This and other criticisms have led to the view that "at the very least, science should evolve into a more applied, interdisciplinary, format for coping with environmental threats, and that it should be seen as a tool for a more open and participatory culture of decision taking" (Jordan and O'Riordan, 1995, p. 61). The precautionary principle, to some degree at least, is meant to shift the onus for the protection of human and environmental health and safety back on to those whose actions are imposing social costs, quite apart from the science aspects that may be involved.

Recent efforts by the European Union to devise guidelines on the application of the precautionary principle are of particular interest (European Commission, 1998; Smith and Halliwell, 1999). These partly derive from EU or community environmental law. The guidelines suggest that the precautionary principle is an approach to risk management and that it applies to risks for future generations as well as to present risks. Because of considerable public opposition to biotechnology in Europe (especially reflected in the European Parliament regarding biotechnology in food), the precautionary principle found its way into several parts of the biosafety protocol.

Following the adoption of the protocol, U.S. officials drew particular attention to why the U.S. believed that the protocol both included the precautionary but also reigned-in on it.

The U.S. Assistant Secretary of State, David Sandalow, first stressed that " the biosafety protocol is the first international agreement to expressly recognize the potential benefits of modern biotechnology" (U.S. Embassy, p. 1). Second, he stressed that the protocol "has ensured ...that world food trade will not be disrupted" (Ibid, p. 2). Thirdly, he stressed that nothing in the protocol changes the requirement under the WTO to regulate "on the basis of sound science" and that the protocol is consistent "with the long-standing U.S. view that the precautionary approach should be part of a science-based decision-making process, not a substitute for that process" (Ibid, p. 2). Quite pointedly, Sandalow also noted that " the protocol does not include a much more stringent requirement sought by the European Union and others that every individual strain or a bioengineered product be identified, in every individual shipment" (Ibid, p. 2).

STATE SUPPORT AND SCIENCE NETWORKS AS REGIME ELEMENTS

International regimes for biotechnology cannot be analytically confined just to rules and dispute settlement processes or to norms such as objective science and the precautionary approach. One must look as well at broader policies of state support (national and international) and at the role of science and scientists in an area where, as we have seen above, science-based governance and controversy is central. In this section, we look very briefly at these broader aspects of a nominal international biotechnology regime in two ways: first, through a discussion of support for aspects of biotechnology through initiatives such as the Human Genome Project; and second, through a look at the mediating role of international science and networks of scientists engaged in risk-benefit choices and advice (Adams, 1995; Smith and Halliwell, 1999).

As mentioned earlier, the political fault lines for the politics of biotechnology is dividing broadly along the food versus health realms. But on the health side (and with inevitable impacts on the larger biotechnology regime) there are several major elements of state support for R&D, especially on initiatives such as the Human Genome Project. Genomics is defined as "a discipline that aims to decipher and understand the entire genetic information content of an organism" (Genome Canada, 1999, p. 7). It differs from classical genetic research in its large scale, broad scope, and heavy reliance on computer-based bioinfomatics but there is little doubt that many see it as the core of a new economy technology (The Wellcome Trust, 1999; Thurow, 1999).

The Human Genome Project is a global publicly funded project whose purpose is to map the human genome or the genetic " book of man". The scale of funding on the human genome project and on other genomics research has been impressive and is largely US and UK led, the latter partly through the Wellcome Trust. The U.S. National Institutes of Health are helping to fund three U.S. Genome Centres but overall U.S. public funding is estimated at over one billion dollars. German, France and Japan have also committed large increases in genome funding (*Nature*, July 8, 1999, p. 102; *Nature*, July 15th, 1999, p. 199 and *Science*, July 9, 1999, p. 183).

But human genome research is also proceeding at breakneck speed in the private sector. And this has resulted in a growing concern about ownership of genomic data. Some of these issues came to a head early in 2000 when negotiations broke down between the Human Genome Project, the publicly funded multinational group, and Celera Genomics, a private US company which was moving even faster than the public project to map the human genome. Other private biotechnology companies feared that Celera would lock up rights under patent protection. Public interest scientists and advocates also feared that it would stifle the free exchange of scientific data and information.

These concerns culminated in a joint statement by President Bill Clinton and Prime Minister Tony Blair issued on March 14, 2000. The two leaders stressed the need for the human genome raw and fundamental data to be kept in the public domain so that scientists could utilize it freely. The Clinton-Blair statement stressed that "unencumbered access to this information will promote discoveries that will reduce the burden of disease, improve health around the world and enhance the quality of life for all humankind" (Quoted in *The Independent*, March 15, 2000, p. 5). But they went on say as well that "intellectual property protection for gene-based inventions will also play an important role in stimulating the development of important new health care products" (Quoted in *The Independent*, March 15, 2000, p.5)

In effect, such rushed leadership statements are a form of moral suasion or exhortation that enter the international regime norms but not necessarily with a clear intent. Confusion immediately arose on financial markets about whether the gist of the speech was to suggest that gene patenting should be banned and as a result Celera's stock value dropped by 21 percent (Alexander, 2000, p. 10). But the speech in fact referred to both free and patented use without specifying which was which. Thus, the exact nature of the international regime for biotechnology remains contested both in market and public policy terms (Reid, 2000).

Much of the pressure which led to the Clinton-Blair statement came from the broader scientific community. Casting the scientific community and networks of scientists as a part of the international regime for biotechnology is quite crucial. The social system of science is quintessential global to start with but a precise mapping of it is difficult when one thinks of it in terms of biotechnology alone. The roles of scientists and their exchanges with each other is multi-faceted and complex and occurs *within and among* all of the institutional arenas implicit in our discussion above: industrial firms; NGOs; universities; several national regulatory bodies; and international institutions. They are engaged in enormously varied networked exchanges about the risks and benefits of biotechnology, the nature of risk management and risk communication, and the technical nature of new products and processes.

One of the larger gathering points for this array of scientific and technical opinion and advocacy was a major international conference in March 2000 in Edinburgh on genetically-modified foods, the most contentious area of biotechnology (*Financial Times*, March 2, 2000, p.11). It brought together about 400 scientists, regulators, consumers representatives, and food industry officials and its work feeds into a further report by the OECD. While the Edinburgh event is only one forum it did bring out two elements that presage further issues for international regime development for biotechnology. First, debate at the conference focused on the issue of "substantial equivalence", the term promoted by the OECD and adopted by regulatory authorities in many countries. This concept required that GM food be as safe as conventional counterparts. Not unexpectedly, anti-GM food advocates argued that genetically-modified food required that a stronger standard of comparison was needed, given what they say are the unknown elements of genetic manipulation. A second point to emerge from the Edinburgh conference was the suggestion from the Conference chair that he would be proposing the "creation of an all-embracing international global forum- inspired by the Intergovernmental Panel on Climate Change– to pull together scientific opinions on GM (food)" (*Financial Times*, March 2, 2000, p. 11.)

THE CORE POLITICS OF THE BIOTECHNOLOGY REGIME

Most of the core politics of the international biotechnology regime has been revealed in the basic account of the four elements of the regime presented above. But it is useful to summarize them in an overall way and to see how they have broadly changed over the last decade as a whole.

First, the United States has been the strongest advocate of most of the trade-related aspects of the regime and of the push to ensure that biotechnology was not singled out as a new technology but rather was governed by generalized trade rules. U.S. political power also was crucial in securing TRIPs and ensuring that intellectual property matters were put in the safer hands of the WTO than in broader bodies such as the World Intellectual Property Organization. Behind this concerted view was the massive lobbying power of the U.S. global pharmaceutical and biotechnology industry but the supportive lobby within the U.S. went beyond these sectors as well since other aspects of IP and trade were also involved (in fields such telecommunications).

The European Union was broadly aligned with the U.S. in the basic thrust of the trade-related and TRIPS aspects of the regime but was much more sceptical of biotechnology in food, in part because this was seen as a part of a U.S. (and North American) trade advantage and because public opinion in Europe strongly questioned the safety of biotechnology in food. As the 1990s ended, the fierce debate on biotechnology in the UK also affected the climate of receptivity for biotechnology in Europe. The controversy over "mad cow" disease also directly affected this climate of debate in the EU and helped generate support for the precautionary approach. Biotechnology in food may also have taken on larger political hues in Europe because it could be linked in key countries such as France to aggressive pressure from the U.S. and elsewhere to open up European agriculture markets and to reduce EU subsidies for agriculture. It could also be linked to broader left-right politics where biotechnology evoked debates about the power of multinationals and, after the December 1999 Seattle protests against the WTO, about perceived unaccountable global governance by the international trade fraternity of policy makers (Lynas, 1999; Hunt, 1999).

The position of developing countries in the various regime development processes were typically defensive. The broad trade-related measures which were nominally not biotechnology specific were negotiated from a position of political weakness. Developing countries signed on to the WTO because other trade-offs were made that were of interest to them. Moreover, they were always faced with the need to insert transition mechanisms to deal with their weaknesses in underlying scientific and technical capacity to implement trade and TRIPS provisions (Trebilcock and Howse, 1995; Lynas, 1999; Doern, 1999). Developing countries continue to have strong concerns about the impact of biotechnology on biodiversity and on issues of the ownership of the technology and its products.

The ascendency of the trade-related aspects of the nominal biotechnology regime arises from the fact that these aspects were constructed and negotiated largely in the late 1980s and early 1990s. But, the latter two elements of the biotechnology regime traced above have emerged more in the late 1990s and into the new Century. Thus the biosafety protocol and even the Human Genome Project reflect a later set of political forces that broaden and deepen the notion of what a biotechnology regime might consist of, or evolve into.

The post-Seattle negotiation of the biosafety protocol may also have caught some of the winds of a changed politics. The protocol emerged out of the biodiversity convention and hence out of an environmental convention that the U.S. had not yet ratified. The U.S. was heavily involved in the biosafety negotiation process (as a part of the Miami Group) but it was not directly at the table. Nonetheless, the U.S. fought hard, as did the Miami group, to ensure that the biosafety protocol could not override trade rules and that the precautionary principle could not become a license to protect, especially an EU license to protect. An effort to square the circle of objective science and the precautionary principle was made but has not been resolved.

But the biosafety protocol is indeed overtly and directly about biotechnology or at least parts of biotechnology. Of some importance here is the fact that in 1999-2000 public opinion in the U.S. was also changing regarding biotechnology. Polling data showed, as in Europe, the UK and Canada, that though there was support for biotechnology in health and in respect to the human genome developments, there was growing concern about GM food and particularly about consumer choice. This latter scepticism had not been present in earlier US polls, partly because GM food was seen as a US trade and industrial advantage and also because American farmers supported GM foods.

Meanwhile, governments had relatively little difficulty justifying their support for the political "high road" cause of human genome research which could be positively linked to future cures for cancer and other diseases. But, the human genome realm did raise the conundrum of who owns the human genome data and hence produced the Clinton-Blair exhortation to do the right thing without defining the right thing.

CONCLUSIONS

This paper has examined the changing nature of the international regime for biotechnology and has also questioned the extent to which an international *regime* actually exists for biotechnology. It has also examined the core international (and domestic) politics which has shaped and established such a regime but which, across the last decade as whole, has also left it in a contested and therefore still evolutionary form. We have emphasized from the outset that the notion of an international regime for biotechnology must be taken as a question because the very problematical nature of international governance. For our purposes a regime has been defined as an interacting set of organizations, statutes, agreements, ideas, interests, and processes engaged in policy development, rule making and implementation in the biotechnology policy field.

The analysis has focussed only on four elements of the notional international regime for biotechnology: the World Trade Organization (WTO) and its dispute-settlement rules overall, its norms about objective science, and the Trade-Related Agreement on Intellectual Property (TRIPs); food and pharmaceutical regulatory approval processes; the biosafety protocol where ideas about the precautionary principle are more in evidence and where biotechnology is an explicit focus; and aspects of state support for biotechnology exemplified by the Human Genome Project and informal and formal channels of discussion among scientists and networks of science.

The analysis has shown that in the first two regime elements, biotechnology can be seen as simply a subset of its rules and norms and indeed key interests and countries led by the U.S. have tried to ensure that it is seen as just another product or technology rather than something that requires special treatment or its own regime. In the third and fourth elements examined, the biosafety protocol and state support activities, we have seen that there is already the notion that biotechnology deserves its own designated "regime" to highlight its differences as an enabling but contentious technology and set of products and processes.

The analysis shows overall that the regime has been contested across the last decade and involves the different and changing core political-economic positions of the U.S., the EU and developing countries. The ascendency of the trade-related aspects of the nominal biotechnology regime arises from the fact that these aspects were constructed and negotiated largely in the late 1980s and early 1990s when U.S. influence was especially dominant but where the EU also agreed with many overall trade aspects of the regime. But, the latter two elements of the biotechnology regime traced in the paper have emerged more in the late 1990s and into the new Century. Accordingly, we conclude that the post-Seattle negotiation of the biosafety protocol may have caught some of the winds of the changed politics of biotechnology. This is a subtle change in political climate whose exact contours are still evolving. Partly, it is centred on changed public opinion in the US about biotechnology in food that has arguably moved somewhat closer to that of the EU, but it also involves public support for biotechnology in the realms of public health where the Human Genome Project shows the more positive side of how people regard biotechnology as an enabling engine for both economy and society in the early 21st Century.

Chapter 11

AUSTRALIAN BIOTECHNOLOGY FIRMS:
Problems in Appropriating Economic Returns to Knowledge?

Maureen McKelvey
Linköping University

INTRODUCTION

One vital issue facing firms and countries is how, and why, they may be able to compete in areas dependent on new knowledge. Currently, everyone seems to agree that new knowledge and innovations are very important to economic and societal change, but few seem to come to agreement about how, or why, the dynamics of these changes actually occur. A number of related concepts such as the 'knowledge economy', 'new economy' or 'learning economy' all try to capture the fact that certain types of knowledge are particularly important to the economy in the present period (OECD 1999).

Some theories focus explicitly on information technology (IT) as a particular explanation for why change is occurring. The 'new economy' debate has many fractions and mass media representations, but the basic argument is that IT has fundamentally changed how the economy works, and therefore, business cycles are a thing of the past. There has been a strong emphasis on the role of small firms, particularly in the American context. Other theories take a broader view about how, and why, knowledge underlies dynamics of the economy. In this latter view, the university and basic research system are seen to play a particularly important role for commercialization (Nelson 1996; MPI 1999). Yet other theories and/or policy advisors suggest a handful of conditions which need to be reproduced in order to recreate 'Silicon Valley, Inc.' anywhere in the world.

This chapter examines how trends in Australian initiatives to commercialize biotechnology relate to different theoretical recommendations. Is the main issue intellectual property rights (IPR)? If not, what other variables help explain the economic dynamics of modern biotechnology and how does that relate to the challenges of small countries? Section 2 reviews three types of theories in order to identify the main factors proposed by each to explain these issues. Different theories propose different explanations as to why firms in certain regions or countries should have relatively greater opportunities, or difficulties, in moving into developing and using new areas of knowledge for economic purposes. For biotechnology, a central characteristic to understand is how and why economic appropriation of new knowledge occurs, in relation to a scientific knowledge infrastructure.

Section 3 then analyzes trends in the Australian initiatives for modern biotechnology, in order to relate developments to theoretical explanations. Which modes of development are currently being followed by Australian policy-makers and universities in order to stimulate the economic appropriation of biotechnology knowledge? The reasons for analyzing Australia are threefold: 1) The national economy is relatively small and traditionally dependent on agriculture and mining/metals. 2) Modern biotechnology is portrayed as one knowledge area in which Australia may be able to successfully compete. 3) The federal and some state governments in Australia have actively pursued policies and initiatives in recent years.

The concluding section will draw out the implications for the specific problems of the economic appropriation of knowledge in small countries as compared to large countries. It draws some conclusions and implications, based on a theoretical and empirical understanding of the dynamics of modern biotechnology. The concluding section thus addresses whether, and why, small countries have choices to make in order to compete.

THEORETICAL PROPOSITIONS ABOUT THE ECONOMIC VALUE OF KNOWLEDGE

Researchers, managers and policy-makers have become increasing aware that (some) new knowledge can be economically valuable. Trying to explain phenomena related to the 'knowledge economy' has stimulated both new theories about the economics of innovation as well as modifications to existing economic theories. Three theories will be briefly reviewed here, especially ones assumed to underlie some of the current government and university policies, both in Australia and internationally. The first theory is neoclassical economics, the second economic geography, and the third is the economics of innovation, drawing from a theoretical informed view of the empirical trends in biotechnology.

Neo-Classical Economics

During the previous century, neoclassical economics formed the dominant body of theories to explain economic processes. The basic idea is that the price mechanism enables comparison among diverse goods and services. In a market, buyers and sellers will buy or sell, respectively, until an equilibrium is reached, depending on their preferences and on their price and income elasticises. An equilibrium is basically a situation where, given the clearing price for the competitive market, no one is willing to buy or sell one more unit.

The conceptualization of the market in neoclassical economists was initially based on standardized, commodified products, where the buyer could choose between identical products on offer from the sellers. It was assumed that the products were undifferentiated and that no one buyer or seller had the power to influence a competitive market. The analysis is backed up by a number of assumptions, which together define this ideal market and how it works. One important foundation for the functioning of the market is ownership and correspondingly, the ability to transfer ownership through market transactions.

For the modern knowledge economy, the challenge, then, has been to modify the basic assumptions and analysis in order to make the theory relevant to the very specific and idiosyncratic characteristics of knowledge for sale in a market and/or used for innovation. Rather than revise the assumptions and explanations, however, neoclassical economists have analyzed knowledge using their traditional logic of interpretation. They use the ideal analysis of the market as something from which to compare the actual workings of a market. Relative to the ideal model, they can then analyze, for example, the role of information asymmetries, habits, and other factors which might 'distort' the market. From there, they make predictions and recommendations about where, for example, that the government can mitigate 'market imperfections', which act to create time lags or other inefficient situations and which thereby 'prevent' the market from working efficiently.

Note, most importantly, that these market imperfections are all ones which lie outside the market, and hence are to some extent outside economic explanation or models proper. Moreover, this analysis implies an assumption of a representative agent, in that there is an assumption in the analysis that all firms would act the same way, given the same set of preferences and stimuli (Hodgson 1988). This makes it difficult to observe, and explain, diversity. Any deviations from the norm have to be analyzed as problems and imperfections, as compared to the ideal situation. This logic of argumentation is applied to new knowledge and innovations, which are basically conceptualized as shocks which arise outside the economic system.

Obviously, the traditional neoclassical analysis does not work so well for the sector of interest here, e.g. rapidly changing, knowledge-intensive sectors. The basic analysis has, therefore, been modified in various ways to apply to these categories of problems. One approach, started already with Nelson (1959) and Arrow (1962), is to analyze the peculiar problems of selling information in a market. For example, unlike goods, information can be used by more than one buyer simultaneously. This facilitates diffusion well beyond the initial producer or buyer of knowledge. This affects broader social development. On the one hand, widespread diffusion may be good for societal development writ large but on the other hand, diffusion may also reduce the incentives of any one firm to invest in developing it (e.g. because others benefit from the firm's investment). Moreover, the buyer has difficulties in evaluating information without having heard it, but once having heard it, the buyer might use it without paying. Note that Arrow (1962) discusses the case of knowledge which can be sold, and not the case of how and why knowledge will be used within the firm.

A second, more recent approach is to acknowledge that information and knowledge are quite important in the economy, and to try to incorporate them somewhat more directly into the analysis. Much of this perspective was stimulated by the early growth accounting literature, which found that a high percentage of economic growth could be attributed to a 'technological residual' (Abromovitz 1989). More recently, the 'new growth theory' postulates the importance of knowledge, for achieving economic growth but again, without changing basic assumptions of neoclassical theory (Romer 1990). This has been mostly interpreted in the debate as human capital, or highly skilled labor.

Neo-classical economics offers few, if any, assumptions about the specific difficulties of small countries competing against large ones - except for the problem of scarce resources. In international trade theory, it should not matter whether firms are in a small or large country. Instead, the country must - and/or is assumed to - specialize into sectors where that countries' firms are relatively more efficient than the others, and then trade internationally. By extrapolation, a small country might have scarce intellectual resources related to size.

Neo-classical economic theory thus provides three explanatory variables about which aspects of knowledge development should be particularly important to create economic change.

Ownership Rights Over Knowledge

The economics of information stresses the importance of ownership over information in order to later sell that information on the market. This implies that firms and countries should be focused on institutions and practices to grant, and modify, intellectual property rights (IPR) that are relevant to biotechnology. The IPR system gives innovators incentives to invest in the development of economically relevant knowledge, and it also provides the means to appropriate economic returns.

Highly Qualified Labour for Knowledge Intensive Sectors

The new economy debate particularly focuses attention on the importance of training, especially of highly skilled labor to be used in firms. As investment and market clearing in education is a long term and less than perfect process, the government may need to direct investment into appropriate skills and training.

Spill-Overs of Knowledge From Public Research To Private Firms

The debate over spill-overs of knowledge is that generating knowledge within one area (of the economy) may have second and third order effects on other parts of the economy. This gives rationale to government investment in the development of general knowledge as a collective good, so that many types of firms may benefit. Moreover, knowledge investments by any one firm may lead to positive stimuli on other firms as well.

Economic Geographers

Economic geographers have their main interest in the relationships between economic processes and geographic or other space through agglomeration. The economic geographers discussed here are ones interested in how agglomeration relates to innovation and economic growth. Some of these also state that they are evolutionary economists, yet they also come from a somewhat different history of debates and have more explicit assumptions about the problems, and possibilities, of space.

Within economic geography, the post-fordism and flexible specialization argument dominated earlier debates about regional development. The basic argument forwarded was that disparate relationships among small firms in a region can be a substitute for consolidating all economic activities within one large firm. The prime example, still largely studied, is regions in (northern) Italy. The flexible specialization debate makes two types of arguments about why regions can experience dynamic growth. One argument related to which types of assets can be shared within a region (such as local labor market, joint relations with distributors, etc). The second relates to what is seen as a more positive form of work relationships and hence a different division of labor across firms, rather than power relations inside one firm.

One important development, particularly in the USA although with European influences, is that economic geographers started to study regions with other types of firms. After all, the Italian regions studied mostly produce commodity goods, such as shoes, other leather goods, furniture, etc. The American studies through, especially, work by Scott and Storper (1992) and Saxenian (1994) looked at regions like Hollywood and Silicon Valley where the firms were involved in movies, electronics, biotechnology, etc. Some of these contributions emphasize the importance of the existing industrial structure and production specialization for influencing the ability of a region to move into a new knowledge area (Maskell et al 1998). Specific benefits of regional agglomeration are also reported, such as specialized division of labor among organizations as well as the need of some 'critical mass'.

Increasingly, however, the debate has moved from the material base of production - and even the geographical base of production – to a social space. Concepts relating to the 'local embeddness' of relationships, the intensity of non-work interaction, the high job mobility, the social worlds of production, and so forth are increasingly used as social interaction explanations about why some regions are relatively more successful than others (Storper and Salais 1997). In other words, what matters is not the economic base of production but the social relationships of interaction. Convention and culture are argued to matter a lot for resulting pathways of development.[1]

Theory in economic geography thus also provides three explanatory variables and may offer more explicit assumptions about the relative advantages, and disadvantages, of small countries as opposed to large ones.

Small Firms Interacting And Local Labor Markets

The debate which started in fordist and flexible specialization indicates that conditions related to labor should be analyzed, such as relative power of workers, local labor markets, as well as conditions about local interactions among firms. The emphasis is on small firms, which exhibit a certain division of tasks as well as local, collective institutions.

Geographical Concentration

The clustering of certain types of activities in physical geography should be evident, although the theories tend to stress relationships among firms, albeit with some discussion of public goods. Two important factors related to business activities are what is actually produced (eg specialization) as well the geographically distribution of firms.

Social Space and Interactions

The more recent turn to the social space implies that the most important explanations should lie with the social types of interactions, including both more formal and informal types of interactions. Relationships, networks, clusters, and other terms stress the social and interactive nature of business as one of the most important explanations for one region's relative success.[2]

Economics of Innovation

The economics of innovation is a broadly defined field of study, with contributions coming from people working in more narrowly defined fields such as science policy studies, economics, evolutionary and institutional economics, and management of innovation (Dosi et al 1988; Tidd et al 1997). Broadly, the field examines the reasons for the development and use of different kinds of knowledge, with contributions ranging from the reasons for success (or failure) of specific firms or sectors at innovation to analyzing the broader knowledge infrastructures in a country. Here, the focus is on theoretically informed work on the specific dynamics of modern biotechnology. This work presents somewhat different variables for understanding how and why knowledge can be used to create economic value.

Biotechnology is one knowledge area, which is important both in terms of scientific understanding and in terms of potential economic potential. Modern biotechnology involves a variety of techniques to gather, use and insert new information in a relatively controlled manner into biological systems in order to fulfill some goal. New information which was never previous found in a species can be introduced. Although the USA in particular has seen the development of dedicated biotech firms, modern biotechnology also affects a wide range of existing sectors. This knowledge area affects many existing sectors such as pharmaceuticals, agriculture, food, and environmental engineering. For economic use, modern biotechnology has its particular impacts in terms of new products, new ways of doing things, and a new way to decrease the firm level search costs underlying innovations (McKelvey 2000).

Modern biotechnology can be compared to traditional biotechnology techniques of brewery or bread, where the same logic of intervention exists but where the techniques and basic knowledge about how things work are less well developed. Moreover, modern biotechnology is more directly and continually related to developments within basic science whereas traditional biotechnology has been based more on long industrial experience, including a trial-and-error method. With traditional biotechnology, the specialized firm relied more on its internal competences and knowledge resources whereas with modern biotechnology, these things are complemented by a more direct influence of scientific research.

In the modern period, there has also been changed patterns of knowledge development and of interactions among organizations. One challenge for initiatives to stimulate the economic uses of modern biotechnology has been the difficulties in conceptualizing and analyzing why the processes of basic scientific understanding and of economic potential sometimes occur in parallel and sometimes intersect. Assumptions still abound that doing high quality research is enough to stimulate economic development or conversely, that doing directed research for firm goals is enough for longer-term economic development. Neither assumption tells the whole story.

Not all scientific results are of economic value nor are all innovations in firms based directly on current scientific results. McKelvey (1996/2000) proposes an analysis based on four differing selection environments, defined in terms of axes of market-government funding and of scientific-technical knowledge. Each selection environment has specific institutions and reward systems to encourage the development of certain types of knowledge. The purpose of that analysis is to help identify and explain some of the conflicts and opportunities which occur when basic science itself has an economic value – even before any 'application' occurs. Basic science is done in parallel with developments within companies. In fact, many researchers have identified the close links between universities and firms as particularly characteristic of this phenomena, although university – government – business relations may be a more general trend in the economy (Etzkowitz and Leysedorf (eds) 1997).

One particularly important aspect of the biotechnology story is that commercialization of the underlying knowledge can lead to complex relationships between different types of firms and between different organizations. These complex relationships may involve the creation of new, specialized biotechnology firms. These firms can follow different development paths. They may try to become independent producing firms in sectors such as pharmaceuticals or environmental technologies. Or, they may become intermediaries between university science and firms in existing sectors in the sense that they develop and sell specific knowledge to existing large firms. Many, however, discuss only the small, biotechnology firms and call them a 'new' industry (Audretsch 2000). However, the small biotechnology firms cannot be analyzed in isolation from their major buyers, e.g. other firms, nor from their knowledge relations (McKelvey forthcoming 2000).

This complexity of relationships is also visible at the level of individuals. Individuals can also have diverse, and complex links with the various types of organizations. Individuals within the research community may spend their time among one or both these types of firms through positions such as owner, board member or scientific advisor and/or they may also spend their time within the university environment. The organizations will, to some extent, specialize in the types of knowledge developed and then rely on market transactions and network interactions in order to gain access to other, complementary knowledge (McKelvey 1997).

This view of the dynamics of modern biotechnology indicates a number of trends which mainly relate to the intersections between the dynamics of knowledge production and the dynamics of market forces. The following are particularly important:

Close Linkages To High Quality, Fundamental Science

Modern biotechnology continues to have strong links to high quality research at universities and research institutes. The reason may partly be the geographical clustering, which is mainly analyzed by economic geographers (above). Another reason, however, is that the quality of that science matters (McKelvey 1997; Audretsch 2000).

Commercialization Of Knowledge Is Visible Through A Variety Of Relationships Among Firms And Knowledge Organizations

Despite the focus on dedicated biotech firms, commercialization of modern biotechnology occurs in different types of firms. These including large firms in existing sectors as well as small dedicated biotech firms. Moreover, the small specialist firms have tended to be dependent on the large firms (although some small do also sell consumer products).

International Knowledge And Market Processes In Local Contexts

Any local or national developments in this field are, at the same time, very much part of an international process, in that both the knowledge and markets are international. They are international although some aspects and modes of development remain national or even local. Anyone wishing to develop and/or use modern biotechnology for economic returns must at the least be aware of international trends in order to stimulate geographically dependent development. Again, this lies close to some of the concerns of economic geographers but the explanations for local as part of the international are here more related to economic and cognitive forces.

Australian Initiatives About Modern Biotechnology

The purpose of this section is to identify Australian trends to match them to the recommendations of the theories analyzed in the previous section. Policy-making often involves, of course, a borrowing of concepts and recommendations from a variety of theoretical perspectives. Depending on the qualifications and interests of the individual and the organization, policy-makers may have anything from little to quite detailed understanding of the underlying arguments for why one course of action might be preferable to another. They often want recommendations for action, not convoluted reasoning about the relative advantages and disadvantages of each possible action. Still, the general trends can be analyzed.

Competing in new, knowledge intensive sectors is a major concern for Australia, despite recent growth. It is a relatively small country (18.5 million inhabitants) but with a large geographic surface (7,692,032 km^2). The majority of Australians live in large cities or medium towns, leaving much of the 'bush' or 'outback' very sparsely inhabited. Although Australians often talk of being a gateway to Asia and the majority of exports go to Asia, the continent (or at least the areas of the continent where people live) is actually a great distance from Asia as well as from Europe and North America.

Australia as a national economy is quite dependent on natural resources such as agriculture and minerals.[3] In terms of exports, agriculture accounts for almost 20% while all types of minerals, metals and fuel account for closer to 35%. This implies that agriculture and mineral/metals/fuel together account for 55% of exports. Meanwhile, manufacturing also accounts for about 20% of exports, but much of manufacturing is, in fact, related to processing food and metals. The Australian figures as to what is actually produced in the national economy show more diversity than these export figures. As a whole for production, services are very important to the Australian economy. Services are 68.9% of production, while manufacturing is 13.3%, mining is 4.6% and agriculture 3.4% (www.disr.gov.au and www.abs.gov.au).

In Australia, the Health and Medical Research Strategic Review, or what is known as the Wills report, addressed the strengths and weaknesses of the Australian health sector (see www.health.gov.au) in the late 1990s. This report seems to have been a crucial turning point, to identify and mobilize support for modern biotechnology. The relevant recommendations here were that the country should invest resources in the basic science related to health biotechnology, in order to stimulate longer-term economic growth. The report's recommendations seem to be based on an assumption that through a few changes, enough incentive structures and high quality research will be available to stimulate commercialization. Applications of biotechnology are hence seen as one area of 'high tech' or 'knowledge intensity' where Australia has the possibility to compete.

Many reports, documents and discussions in Australia have of course centered around these issues after the Wills report. The following analysis is based on the examination of such documentation and of major initiatives from the federal government (departments) such as Department of Industry, Sciences and Resources (ISR) and from three of the major states (New South Wales (NSW), Victoria and Queensland). The purpose is to classify the major forms and motivations of initiatives taken within the Australian context.

As would be expected given current international discussions and policy initiatives, Australia policy and initiatives seem to particularly emphasize some themes rather than others. Analyzing the themes as relatively more or less common does not mean, however, that all federal agencies and state governments immediately and directly supported all initiatives or ideas. Indeed, there has been diversity, with some agencies moving much more quickly than others. The point here, however, is to show the general trends of the Australian initiatives, rather than analyze each one in detail.

The following themes are quite clear:

Importance Of The Ownership Of Intellectual Property Rights

The Australian discussion about IPR involves a variety of discussions and attempts at changing the previous research system to stimulate commercialization. It ranges from encouraging university researchers to take out patents to an awareness of the general possibilities of obtaining revenue through royalties. These may be thought to concentrate in universities / research institutes or else in companies.

Importance Of Basic Research As A Motor, Or Means, For Universities To Spin-Off Small Companies And Thereby Create Economic Development

The focus here is on encouraging university researchers to commercialize their discoveries, through a variety of formal and informal mechanisms, into companies. The University of Queensland in Brisbane, for example, has developed the 'Centre for Drug Design and Development' and 'Institute for Molecular Biosciences', with funding also from the Queensland government and from federal sources. One explicit goal is to create economic development through the start-up of many small companies (UQ 1999).[4]

These first two themes currently crop up in most discussions around the world about how to stimulate economic growth through 'high tech' or 'knowledge' areas.

One theme of the NSW government policy is somewhat less common in the current international policy debate. That is the theme of

Diffusing Information Internationally In Order To Match Potential Buyers And Sellers Of Specialized Knowledge

These initiatives mainly involve stimulating international and longer-term contacts of Australian biotech firms with potential purchasers of that knowledge (e.g. large firms), such as through international biotechnology 'trade fairs' or meetings. In effect, the policy-makers are involved in trying to get the biotechnology firms to participate much more effectively in the international market for specialized knowledge.

Two themes which are currently fairly common in international policy-making circles seem not as explicit in Australia at the moment.

N1) Creating Or Improving Social Networks Among Actors

Although there is an implicit recognition of the importance of clusters and interactions, policy and initiatives seem more directed towards the geographic concentration of activities than on the social space. Locally, however, clusters in a social science may be used to try to stimulate development (See N2 below).

N2) Other Collective Attributes And/Or Services That Should Be Available Provided Collectively (Somehow) But Available To All The Individual Firms

These collective attributes might be things like venture capital, specialist lawyers, etc. The assumption seems to be that explicit high-level policy is not needed or else better supplied through other means. These things might take care of themselves through market 'pull', such that if/when a market exists for such collective services, then this will thereby stimulate new entrants to provide specialized services useful for many. The other alternative is that these will be taken care of at a very local geographical level, such as through a science park or an institutional infrastructure provided by a university. The Australian Technology Park in Redfern (Sydney) is one such attempt, which would explicitly like to develop biotechnology firms.

AUSTRALIAN SITUATION IN BIOTECHNOLOGY

Having examined the trends of current Australian initiatives, this section concentrates on the empirical situation about modern biotechnology in Australia. This analysis, as well as the concluding sections, draws further on the economics of innovation perspective as introduced above. This view argues that interaction between research specialization and firms using and/or developing modern biotechnology are particularly vital. The following analysis should be considered as a first exploration of the Australian situation, largely based on existing material.

Partly due to international debates about national systems of innovation, the relative specialization and orientation of national science policy and of firm R&D are increasingly understood as important variables from which to predict future innovativeness of companies and countries. Australia performs about 2.5% of world research, thereby indicating that 97.5% is done elsewhere. Similar figures of world output apply to Sweden, which has been topping the charts in recent years in terms of percentage of GDP spent on R&D.

Private and public expenditures on R&D are thought of as complementary investments to science policy. In Australia, approximately 1% of GDP goes into business R&D, and this means that Australia spends about half the OECD average of 2% (OECD 1999: page 13). Similarly, innovation surveys from the 1990s show that only a low percentage of Australian firms tend to innovate.

Australia is considered to do better in terms of scientific results for government financed research. The Australian government finances research oriented towards modern biotechnology within expenditures for federally funded research money. Ernst & Young and ISR (1999) calculated the estimations of Australian public R&D spending on biotechnology, as found in Table 1.

Table 1: Australian spending on biotechnology in 1999 (Australian and American dollars; percentage of that source's total funding). Divided by source.

Source	Biotechnology, A$ (millions)	Percentage of that source's total funding	Biotechnology, US$ (millions)
CSIRO	40	5%	25.1
National Health & Medical Research Council	40	19%	25.1
Australian Research Council	35	8%	21.9
Cooperative Research Centres	25	18%	15.7
R&D Start	15	9%	9.4
Rural R&D Corporations	10	7%	6.3
Pharmaceutical Industry Investment Program	2	29%	1.3
Universities	90	8%	56.4
Other	0		
Total	257	8%	160.9

(Ernst & Young 1999; 15 and own calculations to USD, based on February 2000 exchange rates)

Table 1 indicates that the total amount of Australian public money spent on research is estimated to be about US$ 161 million in 1999. Biotechnology gets around 8% of total funding from these public sources. Health and agriculture are, not too surprising, major areas of investment in the area.

Beyond research policy, the type and composition of national firms is important. The economics of modern biotechnology can either be interpreted as a small firm phenomena or as a complex division of labor, where small biotechnology firms exist in symbiosis with universities and large firms in other sectors (McKelvey forthcoming 2000). Both sides will be briefly examined here.

The Ernst & Young and ISR (1999) study of Australian biotechnology presents some figures which compare Australia to Canada and USA, based on the annual E&Y biotechnology comparisons. Table 1 gives some comparative figures, as a first approximation of the Australian situation.

Table 2. Comparing Biotechnology Firms in Australia, Canada and USA; 1998/99. Australian dollars.

	Australia 98/99	Canada 98	USA 98
National labour force	9.2 mill	15.3 mill	136 mill
Core biotech companies	120	282	1,283
Average Revenue (A$m)	8.04	4.41	21.3
Average R&D (A$m)	1.95	2.27	11.4
R&D intensity	24%	53%	53%

(Ernst & Young 1999;18)

Table 2 indicates that Australia should have about 120 biotechnology firms, of which 20 are listed and 100 are privately held and unlisted. It also indicates some similarities and differences among the three countries.[5] In contrast, the NSW government commissioned a study which found only 87 small and medium biotechnology firms in all of Australia in 1999 (NSW 1999b). So, the current estimations are somewhere between 87 and 120 firms.

Table 2 requires some interpretation. It seems to indicate that the average revenue for Australian biotechnology companies is double that of the Canadian firms while spending half as much of sales on R&D. This could be interpreted as Australia having great success with some biotechnology firms growing rapidly.

However, one alternative and more plausible explanation is that while the Canadian figures mainly refer to small, dedicated biotechnology firms, the Australian figures also include the manufacturing and research activities of large, multinational companies, which also engage in many other activities in Australia than biotechnology research. This interpretation seems supported by the breakdown of revenues. Of the 120 Australian firms that E&Y identified, the 20 listed companies have together total revenues of approximately US$460m whereas the 100 private and unlisted companies have together total revenue of approximately A$148m. Thus, a few are quite large and many are quite small. The small or even very small size of most Australian companies is supported by company specific data in a database created by Australian Biotechnology Association Ltd and ISR. Finally, in Australia, the size of foreign-owned firms is generally much larger than Australian-owned firms, especially for sectors with a relatively high R&D intensity (ISR 1999c; 32).[6]

Finally, it is worth looking at the strength of two sectors which will be affected by modern biotechnology, namely health and agriculture.[7] The Australian national accounts show activity in both. Agriculture production in Australia has strong international exports as well as a national research and diffusion base. There are also some food processing companies selling to the domestic (and to some extent the Asian) markets. As for health, public and private expenditures is, of course, put into health care (8.5% of GDP). However, much of the Australian market for medical devices and pharmaceuticals are dominated by American owned firms. Australia has a small domestic pharmaceutical sector, with a few small national companies making post-patent drugs. Australia also has production and some other service oriented subsidiaries for multi-national pharmaceutical companies, but which mainly sell to the Australian market.

One last empirical characteristic which fits with the international pattern is that the Australian biotechnology firms have a strong local concentration, or clustering. The main centers are in specific suburbs of Sydney, Melbourne and Brisbane (NSW 1998, 1999a, 1999c; Ernst & Young and ISR 1999). Not too surprisingly, either, as compared to international patterns, these clusters are located near major research universities.[8]

CONCLUSIONS

This chapter presents current trends in modern biotechnology in Australia through the framework of competing and somewhat contradictory theories about how and why new knowledge has economic value. This concluding section addresses whether, and how, these strategies and empirical patterns indicate particular problems, and opportunities, for a small country like Australia, by continuing to develop the economics of innovation view.

The basic issue being raised by the whole issue of economic appropriation of new knowledge intensive areas is whether, and how, any firm or country can be thought to innovate, relative to its past and present specialization. On the one hand, any existing firm or country can be seen as being relatively more or less specialized in different sectors of the economy. For example, one firm may have been traditionally oriented towards mining exploration whereas another has been in heavy engineering and another is a newly privatized organization which previously provided public health services. What possibilities does each, respectively, have in relation to development and use of new knowledge? One hypothesis is that their innovation will be close to their past and present specialization of production and knowledge (Teece et al 1994). This indicates that countries will exhibit 'path dependency', where past choices influences current and future choices.

The paradox to understand, however, is how new sectors and new areas of knowledge will arise in some manner out of an existing specialization of production and knowledge infrastructure. If they were completely bound by the past, then nothing new could arise. One main question to ask is how renewal can occur. It appears that the process of structural renewal of the economy can take different forms. America, for example, seems to have been particularly adept in the 1990s in developing institutions and arrangements to stimulate the development of new, start-up firms. Other countries like Japan have traditionally relied on existing large firms moving into new areas. This strategy was thought to work very well during the 1980s but not very well during the 1990s.

The economic history of biotechnology shows, however, that rather than facing this dichotomy, novelty and economic appropriation will develop out of complex relationships between universities and firms, between small and large firms, and between large firms. Developing a few small firms is not enough. They have to be part of local and international networks, both of knowledge and of market relations. The challenge for Australia is that on the one hand, some high quality research is nationally located and this gives possibilities to generate and sell knowledge of economic value. To do so, forms of ownership and visibility in the international market are quite important as one of the few means to capitalize economically and internationally from national strengths. One reason that Australian universities and small firms must look international from the start is the relative lack of strong firms in related health sectors. Either domestic firms which are innovative in the world market or else multinationals with strong research and development bases may be necessary in order to stimulate closely related economic growth in the country. For agriculture, the situation seems more hopeful for seeds, although Australia lacks the other triangle leg of modern agriculture, namely chemicals. This assessment of Australian national and international possibilities matters for longer-term use of biotechnology.

There are reasons why knowledge development and economic appropriation of new knowledge are both geographically bounded and international at the same time. Given appropriate organizational and institutional arrangements, some knowledge which is codified can move at a fairly low cost, anywhere in the world – at least among specialists. Other types of knowledge are tacit and dependent on business and social relationships, thereby necessitating geographic concentration. These distinctions between these types of knowledge are extensively analyzed in the literature and will not be pursued here (see Saviotti 1998). The point is to introduce the idea that in order to analyze the effects for small countries, it may be necessary to distinguish between some knowledge which is very geographically and/or organizationally bounded whereas other types of knowledge are more internationally mobile.

In fact, the local nature of knowledge is the flip side of its global dynamics. During the post-World War II period, there has been an understanding of the necessity of performing very high quality local research as a part of the international knowledge community. The question more recently has been how the international basic scientific community relates to the commercialization of knowledge in particular fields. What is interesting in modern biotechnology and similar knowledge areas is that the products (goods or services) where the knowledge has value are also international. For some types of products, the concept of a 'domestic market' has no meaning because from the very start, the value of that knowledge is measured in relation to the value of other knowledge around the world. (That knowledge may be sold, as through R&D contracts and licensing, or may be embodied in machinery or in the performance of services). In contrast, for other types of products, that knowledge is dependent on applying that knowledge to a very peculiar situation and/or context. In that case, there may be a strong need to be locally oriented and have local relationships.

The question still be answered empirically and theoretically is how much of each knowledge is needed in which circumstances! Still, this perspective leads to propositions relevant to when, and why, small countries can also compete in knowledge intensive areas. This gives us two models for economic appropriation of knowledge.

The first resulting proposition is that companies and countries can compete based on two separate models. One model is that the (often scientific, medical and/or engineering) knowledge developed locally must be of high quality from the start in order to be sold at all. The other model is that some subset of the internationally validated knowledge is applied to its local context of use. Local here could apply either to geographic (such as a region) or organizational (such as a firm) context.

The second resulting proposition is that for each of these two models, different types of firms / organizations may be better suited to commercialization of knowledge. The second proposition can be used to identify how commercialization occurs.

The first model assumes that high quality, international research is a necessity in order to create knowledge which may potentially have economic value. Some of that basic research will have economic value, some will not. One resulting hypothesis is that universities and research institutes or else small dedicated biotechnology firms should have the advantage in creating this type of knowledge. Much of the knowledge developed under this model will, however, have a high scientific quality but not an economic value. This could be seen as a higher risk strategy, with the potential either for greater returns or greater failure.

The second model assumes the adaptation of knowledge to specific contexts of use. Still, this process will need to be linked into on-going fundamental science in order to solve continuing, and new, challenges associated with applications but not necessarily as closely. The hypotheses here is that either small dedicated biotech firms oriented towards specific uses or else large firms in existing sectors should be most viable. This second model could be seen as a more incremental strategy based on past strategies. It necessitates quality basic science but the potential of rewards and failure are relatively less than in the first model. Still, in order to succeed, small biotech firms need links – nationally or internationally – to the large firms in existing sectors.

In the Australian case, it appears that the main trends lie towards the first model, with commercialization of high quality research. This is a high-risk strategy, and the emphasis is mostly on IPR and small firms. However, one needs to consider the implications of this argument for future dynamics.

One general implication of this view of the dynamics of interaction among knowledge development and economic appropriation is that the links among the research infrastructure and the current specialization of production and exports runs much deeper as an explanation than intellectual property rights. A second implication is that knowledge and production specialization need to be analyzed in order to indicate which of the two models may be more appropriate – or if other models may be more viable for national conditions.

One problem with the current strategy of encouraging economic development through spin-offs from the universities could well be that it robs the basic science of some of its potential. This matters greatly if the continuation of scientific research and the access to the global scientific community will affect the likelihood of future economic specialization. It depends on how the complex networks of national and international relationships evolve over time.

With all the emphasis on how basic science will generate spin-off companies, discoveries and patents, the discussion seems to neglect the internal dynamics of basic science. There is little discussion of why basic science may be useful in its own right, when following its own rules and dynamics. These are situations where the explicit goals of commercialization and further financing the researcher (or university) would only come somewhere further down the line, after high quality basic science has been done.

The development of, and economic appropriation from, new knowledge is thought to be crucial today for competitiveness. Therefore, if knowledge matters, it also matters what research is financed, and performed, by firms and by national governments within a geographically bounded nation. Not only does it matter what is produced, the argument has been advanced above that fundamental research of the highest international quality is one of the necessary ingredients for commercialization. Therefore, even to stimulate the goal of economic development, some of the internal dynamics of fundamental science need to be respected and further supported.

Chapter 12

COMPETING BUSINESS MODELS IN THE FRENCH BIOTECH INDUSTRY

Vincent Mangematin
INRA/SERD, Université Pierre Mendès

INTRODUCTION

Public authorities have recently started supporting development of the biotechnology sector by encouraging start-ups and creating favourable environments such as incubators, a specialised stock exchange or set of technopoles. The different programmes used to encourage biotech development (P. Monsan, 1999) (subsidies for research performed jointly by firms and academic labs, subsidies for start-ups, creation of incubators) seem to be successful if the results are estimated in terms of the number of new firms (around 300 SMEs still in existence, since 1990).[1] On 1 January 1999 France had just over 400 biotechnology SMEs employing a total of 15,000 people, with an estimated turnover of 2 billion euros.[2] Estimates based on the survey initiated by the MENRT[3] are consistent with information published by Ernst and Young, although they indicate a higher number of firms in France. Average size in terms of number of employees per firm is nevertheless similar: about 40 persons, compared to about 140 in USA. All in all, biotechnology remains a small emergent sector compared to others such as agri-food (over 4,200 French firms with 372,300 employees and a turnover of 100 billion euros) or pharmaceuticals (94,500 employees in 271 firms and a turnover of 28,5 billion euros).[4]

The creation of many start-ups during the past ten years raises questions on the future of biotech SMEs in France and in Europe. How will consolidation of the sector occur? Will maturity of the sector be accompanied by progressive disappearance of SMEs and the growth of firms? Will it be structured like the automobile industry, around collaboration between firms with a high capacity for integration of research performed elsewhere, and specialized firms such as parts manufacturers? What will the future be of the hundreds of small firms which focus on the local market or a specific technology, and which have grown at a moderate pace over the past ten years?

To answer these questions we have chosen to reason in terms of business models, that is, organisational models covering the targeted market (final or intermediate market), networks of partners, and shareholders. A business model corresponds to a set of key resources for the firm's development and to a mode of securing these resources within the organisation. It provides insight into the development logic of biotechnology SMEs in France.

The aim of this chapter, based on the results of a 1999 survey on all biotech SMEs in France is to understand the development logics of these firms. The first part defines the concept of a business model. The second part is an overview of biotech firms in France and their development, based of the survey carried out in 1999. It presents three business models for biotech SMEs. The third part is an attempt to map out the development trajectories of SMEs and the respective leading forces in each type of firm. Concluding remarks present three possible scenarios for the evolution of the sector.

ORGANISATION OF FIRMS AND BUSINESS MODELS

Business Model as an Archetype

The formal and informal structures of firms and their external linkages have an important bearing on the firm's growth rate. According to Teece (D. J. Teece, 1996), firms' distinctive modes of governance depend primarily on their boundaries, formal internal structure, informal structure and external relations. Rather than specifying all possible combinations, we have chosen to reason in terms of archetypes.

Greenwood and Hinings (R. Greenwood and C. R. Hinings, 1993) define an archetype by two general statements: "First, organizational structures and management are best understood by analysis of overall patterns rather than by analysis of narrowly drawn sets of organizational properties. [....] Second, patterns are a function of the ideas, beliefs, and values – the components of an interpretive scheme – that underpin and are embodied in organizational structures and systems", (p 1052).

Research on archetypes generally focuses on large firms; only one of the archetypes identified by Teece (1996) describes a small organisation (the individual inventor and the stand-alone laboratory). Greeenwood and Hinings contrast bureaucratic organisations and professional organisations. These typologies are multidimensional. Generally, they are based on organisational structures (rule-making, formal structures of authority, etc.) or on decision-making processes. Yet a range of writings reflects this view of archetype as configured structures expressing underlying values. Ranson, Hinings and Greenwood (S. Ranson, C. R. Hinings and R. Greenwood, 1980) emphasise that it is necessary to investigate the social mechanisms which determine the structuring process and shape the ensuing structural forms if we are to fully understand the formation of organisational structures. One of the dimensions that has been identified to study architectural practices is motives for work, i.e. an intellectual ethos, a set of ideas about the architecture, that results in a particular set of organisational arrangements. A similar idea has been developed in Karpik's conception of a 'logic of action' and in Callon and Vignolle's notion of 'forms of coherence' (M. Callon and J. Vignolle, 1976).

High-tech SMEs are created around a sound project supported by the creators. This project strongly structures the resources and competencies that the firms will have to mobilize.

Mobilising Resources to Innovate

While research on strategy has traditionally focused on an analysis of competition (M. Porter, 1980), analysis in terms of resources and competencies has been developing since the mid-1980s. This change reflects the shift of interest from external towards internal analysis: organisations are studied from within rather than in relation to their environment. The increasing openness of organisations has gone hand in hand with a relative disappearance of boundaries. Thus, the theory of resources, for which the definition of boundaries is less fundamental than it is for approaches focused on competition, seems particularly rich. It proposes an analysis of the organisation and of its competitive advantages in terms of tangible and intangible resources and competencies (J. B. Barney, 1991; J. B. Barney, J. C. Spender and T. Reve, 1994; R. Grant, 1991; G. Hamel, 1991). Competitive advantage is based on a logic of comparative advantages derived from resources and competencies. It is by knowing and controlling them that strategic options can be defined and a competitive advantage created.

Resource-based theory (E. Penrose, 1959; M. Peteraf, 1993; J. L. Arrègle, 1996; J. Mahoney and J. Rajendran Pandrian, 1992; P. Shrivastava, A. S. Huff and J. E. Dutton, 1994; M. V. Russo and P. A. Fouts, 1997) like the theory of dependent resources (J. Pfeffer and G. R. Salancik, 1978; P. S. Tolbert, 1985; A. Valette, 1994; K. Weick, 1979) distinguishes resources which are inputs into the production process and can be of various kinds (capital, human resources, equipment, cooperative networks, or commercialisation networks, reputation or scientific visibility), on the one hand, and competencies which are related to the use and implementation of those resources, on the other. The sustainable nature of competitive advantage depends on the difficulty another organisation would have imitating the source of the reference organisation's success.

This approach seems particularly fertile for analysing an industry that relies primarily on a combination of resources because few of its products have as yet been marketed. Studies in organisation theory are based on a logic of supply. The identification of critical resources for each organisation helps to understand logics of cooperation and to contribute towards the analysis of modes of inter-organisation coordination in the context of resource-based theory.

In the biotechnology sector, several authors have analysed firms' strategies for acquiring and stabilising the resources needed for their business and growth. G and A Eliasson (G. Eliasson, Eliasson, A., 1996) point out the differences of the biotechnology sector compared to other industrial sectors:

(1) Biotechnology is a sector that stems from academic research. Technically, the differences between an academic and a private laboratory are small. However, competencies involved in implementation and industrialisation differ substantially.
(2) The main costs are those of research and commercialisation; production costs are relatively low.
(3) As Henderson *et al.* (R. Henderson, L. Orsenigo and G. Pisano, 1999) point out, discoveries result from the combination of different corpuses of knowledge and know-how. G. and A. Eliasson define all the competencies necessary for the discovery of new products or processes as a "competence bloc".
(4) Finally, when the competencies needed to innovate are scattered among several organisations, agreements between public and private organisations become essential.

G. and A. Eliasson show that the notion of a competence bloc exceeds purely scientific competencies: "to have a business potential, a competence bloc requires a minimum of more or less related competencies embodied in active, competent and resourceful consumers, innovators who select innovations that satisfy economic criteria, competent venture capitalists who recognize and finance commercially viable opportunities, and industrialists". These elements are added to the ability to form partnerships with laboratories or firms with complementary resources.

To understand the logics underpinning the creation and development of biotechnology firms in France, the firms in our sample can be positioned in relation to the different dimensions defined by G. and A. Eliasson:

- Following the article by A. Nilsson (A. Nilsson, 2000), two models of firms are identified. In the first, firms have integrated all functions, including commercialisation and marketing. These firms cater for final consumers, to whom they supply agri-food, cosmetic or pharmaceutical products. In the second model, biotech firms supply intermediary products which are integrated into the production processes of firms directly in contact with consumers. In our study, consumers' competencies were identified in this way.

- Industrialists and venture capitalists have a similar role. They help to finance firms by providing capital. They also link the firm, initially based on a scientific idea, to the business world. This connection has several forms: contact with potential industrial or commercial partners; opening of a market, particularly that of the parent company or other firms in the group, advice and support in strategic management (see *Nature Biotechnol* N°17, Supplement on Bioentrepreneurship, May 1999).
- Liebeskind *et al.* (J. P. Liebeskind *et al.*, 1996) describe the environment of biotechnology SMEs as hyper-competitive. In this context, the appropriability of research results in central. Innovation opportunities are grabbed particularly fast (D. Teece, 1986) when the appropriation regime is strong. International patent laws stipulate that only the first to discover a product or process can take advantage of the discovery (if it is patented). Thus, biotech SMEs compete with established firms which finance their own research – as Arora and Gambardella (A. Arora and A. Gambardella, 1990) show –, with other organisations engaged in biotech research (SMEs and universities) and with potential entrants.

  Zucker *et al.* (L. Zucker, M. R. Darby and J. Armstrong, 1994) show that 97% of star scientists work in universities or non-profit research institutes. Only 3% work in enterprise.
- Even though star scientists do not work directly in biotech firms, these firms do have strong links with the academic world. Dedicated Biotech Firms (DBFs) are thus forced to develop organisational arrangements that give them access to external intellectual resources. The essential character of external resources is threefold:

  (1) Relations with the academic world enables firms to explore a wide variety of hypotheses while maintaining a large degree of flexibility. Thus, exploration of scientific hypotheses in other organisations enables them to reduce costs and to avoid sunk costs, but still to be the first to benefit from the discovery.

  (2) Powell (W. W. Powell, 1990) argues that social networks are the most efficient organisational arrangement for sourcing information because information is difficult to price (in a market) and to communicate through a hierarchical structure. Social networks serve as sources of reliable information, which is essential to efficient organisational learning. When knowledge is distributed among several organisations, not only access to information but also learning how to work in partnership become key variables in competition (W. W. Powell, Koput, K.W., Smith-Doerr, L.,

1996; B. L. Simonin, 1997).
(3) Relations with the academic world, via the social networks in which the creators of the firm are involved, enable the firm to have access to unique scientific expertise that is a critical resource for its survival and development.

To sum up, the hypothesis tested in this article is that French biotech SMEs can be described in terms of a limited number of business models structured by the way in which these DBFs mobilise competencies in order to innovate.

DATA AND METHODS

Biotechnology: A Small Industrial Sector

The French biotechnology sector is an emergent sector consisting of about 400 small businesses in widely diverse markets. 221 firms responded to our survey conducted in the first half of 1999. 186 complete answers were processed. (See box for the detailed methodology of the survey.)

Firms that do customised work for companies in direct contact with consumers

Like everywhere else, French DBFs have little contact with the end user. Only 12% of firms have such contact (mainly for agri-food products, cosmetics and, to a lesser degree, human health). 88% of biotech firms are active suppliers of intermediary goods and services for other firms in the fields of human or animal health, cosmetics, environment or agri-food.

As shown in Table 1, the majority of biotech SMEs specialise in the design, development and production of customised genetic or biological material. A small minority (20%) is engaged in product development (human health – 8% and other, apart from drugs – 12%). Most firms are service providers. They sell biological material to companies in all sectors, enabling them to produce either more quickly or better quality products (e.g. more standardised and better controlled quality), at a lower cost. 48% of firms produce goods, 18% supply products and services and 34% provide services only.

Table 1: Activities of Biotechnology SMEs

			Speciality				Total
			Product development	Development aid or sequencing methods	Diagnosis and manufacturing of biological material	Equipment and material	
Sectors Customers	agri	N° empl.	17	1	27	3	48
		%	35.4%	2.1%	56.3%	6.3%	100.0%
	agri/cosmet.	N° empl.	1		1		2
		%	50,0%		50,0%		100.0%
	agri/cosmet./pharma.	N° empl.	2	3	15	2	22
		%	9.1%	13.6%	68.2%	9.1%	100.0%
	Cosmet.	N° empl.	4	1	2		7
		%	57.1%	14.3%	28.6%		100.0%
	Pharma.	N° empl.	13	20	32	3	68
		%	19.1%	29.4%	47.1%	4.4%	100.0%
	All	N° empl.		6	25	8	39
		%		15.4%	64.1%	20.5%	100.0%
Total		N° empl.	37	31	102	16	186
		%	19.9%	16.7%	54.8%	8.6%	100.0%

Khi-deux Tests

	Value	ddl	Asymptomatic signification (bilateral)
Pearson's Khi-deux	45.838	15	.000
Probability	53.148	15	.000
N° of valid observations	186		

a 13 cells (54.2%) have a theoretical number of employees of less than 5. The minimum theoretical number of employees is .17.

The activity of biotechnology firms is described in relation to two dimensions: the business sector in which the firm sells its products or services, and the company's core competencies (or speciality). The sectors in which it sells its products or services are described in the following way:

- The firm caters for a sector: agriculture or the agri-food industries (26%), cosmetics (4%) or pharmaceuticals (37%).
- The firm commercialises its products and services in two sectors which do not require a marketing license: agriculture/agri-food and cosmetics (1%).
- The firm generates a cash flow from the sale of products and services in the cosmetic or agri-food sector and carries out research in the human health sector – a sector in which it takes longer to generate turnover (12%).
- The firm has generic know-how which it uses in all sectors (equipment, materials, etc.) (21%).

The core competencies of the firm describes the products and services that it designs, produces and markets. Four categories have been identified:

- Product development (20%). The firm's business is production and marketing of products. It does not produce customised products only; it also mass-produces.
- Diagnosis and creation of tests and/or biological material (55%). These firms develop two complementary activities: a) as service providers to other companies they create tests, biological material with specific characteristics, and customised diagnoses; and b) they design, produce and commercialise diagnostic kits, either directly or through other companies.
- Design and production of equipment and material for laboratories. These firms cater for all sectors. They account for 9% of the total.
- Development aid methods or sequencing. These firms design methods enabling firms to improve their processes or to market their products more effectively (e.g. CRO).[5] These firms, which account for 17% of the total, cater primarily for the pharmaceutical and agri-food sectors.

To sum up, out of the firms active in the human health sector, few are directly engaged in the production of drugs. Firms in the agriculture, agri-food or environment fields produce mainly seeds or foods with specific characteristics (health or functional food). Service firms in these sectors mainly provide tests or diagnostic kits for agriculture or agri-food and the environment. SMEs focused on the cosmetic or animal health sectors have largely the same characteristics: close to firms in the human health field, they develop products or services which do not require specific marketing licenses. This market positioning often corresponds to a strategy for progressively conquering the human health market.

Firms active in all the sectors are mainly those which design and develop generic tools or methods (such as sequencing or instrumentation).

Mainly New DBFs

Close to 70% of firms which were still alive in 1999 had been founded after 1990. Firms that are 20 years old or more now account for 12% of the sample, while the most recent firms account for 69% of the total. Less than 20% of all firms were created between 1980 and 1990.

Table 2: Size and age of biotechnology firms

			Age			Total
			Old	Med	Recent	
Customer sector	Agri	N° of empl.	9	10	29	48
		%	19%	21%	60%	100%
	Agri/cosmet	N° of empl.			2	2
		%			100%	100%
	Agri/cosmet/pharma	N° of empl.	1	5	16	22
		%	5%	23%	73%	100%
	Cosmet	N° of empl.	1	1	5	7
		%	14%	14%	71%	100%
	Pharma	N° of empl.	6	13	49	68
		%	9%	19%	72%	100%
	All	N° of empl.	3	8	28	39
		%	8%	21%	72%	100%
Total		N° of empl.	20	37	129	186
		%	11%	20%	69%	100%

Khi-deux : Value 6,131, ddl : 10, significance : 0,80

Average size of firms in number of employees

		AGE			
Customer sector	Customer sector	Old	Med	Recent	Total
	Agri	99	28	30	42
	Agri/cosmet			8	8
	Agri/cosmet/pharma	98	12	20	22
	Cosmet	110	36	4	39
	Pharma	150	64	29	53
	All	20	22	11	14
Total		108	37	23	37

The average number of employees of an DBF is 37 persons, for an average turnover of 480 KE.[6] Whether in terms of turnover or number of employees, these firms remain small. 72% have a turnover under 1.5 ME,[7] compared to only 4% with over 15 ME. 24% have a turnover between 1.5 and 15 ME. Moreover, 55% employ fewer than 10 persons and only 14% employ over 50 persons. 31% employ between 10 and 50 persons. Although a slight difference is apparent between recruitment and generation of revenue, at the time of start up, the number of employees and the sale of products and services remain very closely linked. Firms established before 1980 and still alive in 1999 grew very fast, especially those active in the human health sector. Comparatively, the firms created between 1980 and 1990 grew more slowly.

Firms active in the agro-food or agronomic sectors and those which are active in the pharmaceutical sector are larger than the others in terms of number of employees. Business creation was spread out over time, and across all sectors equally.

Business Models

To understand the growth of firms and the dynamics in which they are engaged, we chose to reason in terms of business models. A business model describes a category of firm in relation to the market it targets, its expected growth and the organisation of its activity. The biotech firms which responded to our survey are spread out in the following way, in terms of criteria borrowed from Eliasson:

Table 3: Distribution of firms, according to the dimensions of G. and A. Eliasson

	Firms which target the final consumer	Firms which produce intermediary products or services
Intermediary firm or integrated firm	12.4%	87.6%

	Venture capital firms	Only natural persons	Other companies	Shareholders other than those listed	Total
Firms whose shareholders consist of	28.5%	32.8%	33.9%	4.8%	100.0 %

	No patent	At least one patent	Total
Mode of management of industrial property	65.6%	34.4%	100.0%

	No partnerships	At least one partnership	*Of which with public labs*	*Of which with universities*	*Of which with foreign labs*
Firm's capacity to form academic partnerships	40.9%	59.1%	*40.3%*	*31.7%*	*6.5%*

The four dimensions proposed, following G. and A. Eliasson, enable us to define three business models structured around management of industrial property, networks of shareholders, and the firm's position in the market.

Table 4 presents a matrix of correlation between the different indicators (shareholders, final or intermediary market, customer sector, patent, age of firm, turnover and number of employees).

Table 4.

		Sh pers	sh vent. K	Sh firm	final cons.	Product	Production standard	Agri ou env	Pharma	Cosm/ veto	All sectors	Hold patent	Creation before 1980	Creation 1980-90	Creation after 90	co-op Labo publics
Sh pers	Corr	1,000														
	Sig.	,														
Sh vent. K	Corr	,226**	1,000													
	Sig.	,002	,													
Sh firm	Corr	-,46**	-,089	1,000												
	Sig.	,000	,229	,												
final cons.	Corr	-,130	,052	,086	1,000											
	Sig.	,077	,478	,241	,											
Product	Corr	,025	,099	,086	,269**	1,000										
	Sig.	,737	,177	,241	,000	,										
Production	Corr	,002	,026	,087	,323**	,732**	1,000									
Standard	Sig.	,978	,728	,239	,000	,000	,									
Agri/envt	Corr	-,147*	-,200**	,098	,145*	-,064	-,004	1,000								
	Sig.	,045	,006	,181	,048	,386	,952	,								
Pharma	Corr	,098	,111	,036	-,131	-,133	-,149*	-,413**	1,000							
	Sig.	,184	,133	,628	,075	,070	,043	,000	,							
Cosm ou veto	Corr	,115	,023	-,064	,197**	,091	,106	-,129	-,003	1,000						
	Sig.	,117	,755	,388	,007	,215	,149	,080	,965	,						
All sectors	Corr	,039	,055	-,052	-,193**	,062	,038	-,369**	-,493**	-,174*	1,000					
	Sig.	,598	,454	,481	,008	,403	,603	,000	,000	,018	,					

		sh pers	sh vent. K	Sh firm	final cons.	Product	Production standard	Agri ou env	Pharma	Cosm/ veto	All sectors	Hold patent	Creation before 1980	Creation 1980-90	Creation after 90	co-op Labo publics
Hold	Corr	,032	,295**	,112	,106	,207**	,142	-,040	,009	,129	-,039	1,000				
Patent	Sig.	,668	,000	,129	,150	,004	,053	,587	,908	,078	,593	,				
Creation	Corr	-,174*	-,027	,065	,186*	,138	,155*	,082	-,089	-,002	-,051	,041	1,000			
Before 80	Sig.	,018	,716	,381	,011	,060	,035	,268	,226	,973	,490	,580	,			
Creation	Corr	-,037	-,046	,079	-,064	,015	,068	-,015	,008	,010	,008	-,021	-,173*	1,000		
1980-90	Sig.	,620	,533	,284	,382	,837	,359	,837	,914	,894	,914	,779	,018	,		
Creation	Corr	,148*	,058	-,112	-,069	-,106	-,163*	-,042	,053	-,007	,027	-,010	-,522**	-,750**	1,000	
After 90	Sig.	,043	,432	,129	,348	,149	,026	,572	,472	,926	,712	,898	,000	,000	,	
co-op avec	Corr	,115	,161	-,155*	-,086	,122	,045	-,075	-,036	-,009	,079	,096	,041	-,134	,088	1,000
Labos publics	Sig.	,118	,028	,035	,241	,099	,544	,307	,628	,908	,285	,195	,575	,069	,232	,

** Significance of the correlation at the level 0.01 (bilatéral).
* Significance of the correlation at the level 0.05 (bilatéral).

Table 5 : Characteristics of each business model

		A	B	C	R	sh pers	Sh vent. K	sh firm	final	Product	standard	Agri envt	Pharma	cosm ou veto	All sectors	Hold patent	co-op labs	INRA	CNRS	INSERM	UNI
A	Corr	1,000																			
	Sig.	,																			
B	Corr	-,349**	1,000																		
	Sig.	,000	,																		
C	Corr	-,211**	-,383**	1,000																	
	Sig.	,004	,000	,																	
R	Corr	-,262**	-,475**	-,288**	1,000																
	Sig.	,000	,000	,000	,																
sh pers	Corr	,142	,162	-,499	,145	1,000															
	Sig.	,053	,027	,000	,048	,															
sh Krisk	Corr	,371**	-,282**	-,212**	,190**	,226**	1,000														
	Sig.	,000	,000	,004	,009	,002	,														
sh soc	Corr	-,127	-,234**	,467**	-,050	-,467**	-,089	1,000													
	Sig.	,085	,001	,000	,496	,000	,229	,													
final cons.	Corr	-,032	,137	,070	-,188*	-,130	,052	,086	1,000												
	Sig.	,669	,062	,343	,010	,077	,478	,241	,												
Product	Corr	,005	,009	,054	-,062	,025	,099	,086	,269**	1,000											
	Sig.	,946	,903	,465	,401	,737	,177	,241	,000	,											
Standard	Corr	,006	,068	,098	-,167*	,002	,026	,087	,323**	,732**	1,000										
	Sig.	,939	,358	,183	,023	,978	,728	,239	,000	,000	,										
Agri / envt	Corr	-,159*	,201**	,150*	-,222**	-,147*	-,200**	,098	,145*	-,064	-,004	1,000									
	Sig.	,030	,006	,042	,002	,045	,006	,181	,048	,386	,952	,									
Pharma	Corr	,136	-,231**	-,103	,233**	,098	,111	,036	-,131	-,133	-,149*	-,413**	1,000								
	Sig.	,064	,002	,161	,001	,184	,133	,628	,075	,070	,043	,000	,								
Cosm / Veto	Corr	,045	,169	-,072	-,161*	,115	,023	-,064	,197**	,091	,106	-,129	-,003	1,000							
	Sig.	,541	,021	,332	,028	,117	,755	,388	,007	,215	,149	,080	,965	,							

		A	B	C	R	sh pers	Sh vent. K	sh firm	final	Product	standard	Agri envt	Pharma	cosm ou veto	All sectors	Hold patent	co-op labs	INRA	CNRS	INSERM	UNI
All sectors	Corr	,025	-,057	-,011	,052	,039	,055	-,052	-,193**	,062	,038	-,369**	-,493**	-,174*	1,000						
	Sig.	,730	,441	,877	,483	,598	,454	,481	,008	,403	,603	,000	,000	,018	.						
Patent Détenu	Corr	,267**	-,134	,086	-,151*	,032	,295**	,112	,106	,207**	,142	-,040	,009	,129	-,039	1,000					
	Sig.	,000	,068	,245	,040	,668	,000	,129	,150	,004	,053	,587	,908	,078	,593	.					
co-op labs	Corr	,067	,009	-,076	,001	,115	,161*	-,155*	-,086	,122	,045	-,075	-,036	-,009	,079	,096	1,000				
	Sig.	,363	,899	,306	,994	,118	,028	,035	,241	,099	,544	,307	,628	,908	,285	,195	.				
INRA	Corr	-,139	-,014	,075	,065	-,014	-,023	-,032	,077	,037	,024	,124	-,181*	-,089	,050	-,009	,343**	1,000			
	Sig.	,058	,848	,309	,375	,847	,751	,664	,296	,617	,748	,091	,014	,229	,496	,899	,000	.			
CNRS	Corr	,071	-,041	-,074	,051	,005	,186*	,027	-,085	,025	-,025	-,206**	,105	-,060	,115	,210**	,379**	,095	1,000		
	Sig.	,334	,582	,318	,491	,950	,011	,717	,250	,732	,739	,005	,153	,418	,118	,004	,000	,196	.		
INSERM	Corr	,134	-,129	-,131	,147*	,070	,254**	,005	-,071	,026	-,038	-,245**	,360**	,002	-,149*	,219**	,357**	,033	,197**	1,000	
	Sig.	,068	,080	,075	,046	,342	,000	,951	,333	,727	,607	,001	,000	,980	,043	,003	,000	,652	,007	.	
UNI	Corr	,109	,028	-,033	-,093	,092	,030	-,144	-,045	,024	-,048	,025	-,028	,037	,018	-,007	,567**	,014	,148*	-,038	1,000
	Sig.	,137	,709	,659	,207	,211	,680	,051	,538	,745	,513	,737	,700	,615	,809	,921	,000	,847	,043	,605	.

** Significance of the correlation at the level 0.01 (bilatéral).

- Significance of the correlation at the level 0.05 (bilatéral).

We can thus identify four groups of firm, the characteristics of which are given in Tables 5 and 6:

Table 6 : The development of each business model

		A	B	C	R
CA nul	Corr	-,143*	-,298**	-,177	,606**
	Sig.	,052	,000	,015	,000
TO less than 1,5 Meuros	Corr	,062	,237**	-,060	-,261**
	Sig.	,397	,001	,414	,000
TO between 1,5 and 15 Meuros	Corr	,055	,014	,250**	-,283**
	Sig.	,455	,851	,001	,000
TO up to 15 Meuros	Corr	,051	,049	,034	-,127
	Sig.	,488	,505	,650	,085
Number of employees Less than 10	Corr	-,072	-,048	-,267**	,350**
	Sig.	,331	,513	,000	,000
Number of employees Between 10 and 50	Corr	,031	,025	,185**	-,218**
	Sig.	,674	,733	,011	,003
Number of employees Up to 50	Corr	,061	,036	,144*	-,218**
	Sig.	,411	,627	,051	,003

** Significance of the correlation at the level 0.01 (bilatéral).
* Significance of the correlation at the level 0.05 (bilatéral).

Group A: Firms With Good Development Potential

These firms vary in size. Their main characteristics are that they patent and are financed by venture capital and shareholders who are natural persons. Considered as the flagship of the French biotech industry, the 30 firms in this group have experienced rapid growth. They specialise in customised production and service provision for pharmaceutical companies. They are rarely presented in sectors related to agriculture or agri-food. 28 out of 30 were created after 1980. Close to 50% of these firms employ over 10 people and a third have a turnover greater than 1.5 ME.

Development of these firms was based mainly on the presence of national and international capital investors. Present from the start, venture capital companies played an active part in their orientation and put them on a fast-growth trajectory. The widening of the circle of shareholders beyond the family circle enabled these companies to benefit from advice, contacts, human capital and an introduction into networks they hardly knew or that would have taken time to discover (under 30% of the 186 firms have a venture capital company as a shareholder). Quotation on the stock exchange enabled venture capitalists subsequently to sell their shares and to withdraw – a condition *sine qua non* for the perpetuation of financing of biotech firms by venture capital. The presence of venture capital firms in these companies paved the way for a transition from an essentially domestic environment (family capital, network of new entrepreneurs) to a truly entrepreneurial environment.

Fast growing firms rely on partnerships with French or foreign universities, as well as with public institutions such as the CNRS or INSERM, to maintain their scientific and technological competencies. For these firms, proximity to a pole of scientific excellence is essential if they are to benefit from spill-overs from public research, through interpersonal relations and international interaction of academic laboratories. To attract the best researchers, PhDs and post-docs – all potential partners – these firms have an interest in taking advantage of centres of academic excellence both at start up and during their development. Several authors (D. Audretsch and P. Stephan, 1996; M. Feldman, 1999; J. P. Liebeskind *et al.*, 1996) have highlighted the role of proximity between high quality academic research and DBFs. Even if these firms are developing collaborative researches with academic labs wherever they are, these DBFs are located near the main pole of excellence, Paris and Strasbourg. Partnerships with other firms, especially pharmaceutical firms, enable them to transfer their technologies. Given the fact that these firms are essentially providers of goods or services to pharmaceutical companies, their turnover is dependent on formal collaboration, as shown by Sharp and Greis *et al.* (N. P. Greis, M. D. Dibner and A. S. Bean, 1994; W. W. Powell, 1998; M. Sharp, S. J. and I. Galimberti, 1994). These contracts are one of the ways of getting round barriers to entry, on both a scientific and a commercial level. They also enable these firms to benefit from faster learning and to acquire additional competencies.

Group B: Firms Which Develop In Niches

These firms all started with the idea of an entrepreneur who mobilised family capital (60% of the firms have only natural persons as shareholders). Very few of them have venture capital firms or other firms among their shareholders. They have little capital and generate turnover from the outset, especially through customised production or service provision. Their turnover is substantially lower (under 1.5 ME) than that of other types of firm. They rarely develop an activity which caters specifically for pharmaceutical companies. On the other hand, they have a strong presence in the agriculture and agri-food fields. They rarely patent their technology although they sell their products to final consumers. These firms look in their immediate environment – family and geographic – for catching the resources to survive and grow. Often they have very few links to networks of actors active in biotechnology (consulting firms, venture capital firms, academics or recognised researchers, ANVAR, ministries, etc.). Relations with the market and users of their products and services are formed on the basis of geographic proximity. They grow as they expand their clientele beyond the local market, by broadening the range of their services or specialising their supply so that they become the leader in their market.

Group C: Firms Attached To A Group

The capital of firms in this group is mainly in the hands of natural persons and other firms. Venture capital is seldom present in firms in this group.[8] Either they were bought out when they were independent, or they were created directly by their parent company. In fact, one of the strategies of pharmaceutical or seed companies is to create biotech firms, either alone or in partnership with other firms. Biotechnologies are a high-risk business whose development relies on specific competencies that are sometimes difficult to maintain within a group. Small structures are more flexible and adjust better than large ones to changes induced by the production of new scientific knowledge. Finally, investing in a biotech subsidiary also enables firms to set up in countries where they can take advantage of externalities of its research and of new markets. Yet firms in this group form fewer ties with academic research than others, especially with INSERM and the CNRS. It seems that the parent company or shareholding company is a special partner, including for research.

Thus, major French companies (Limagrain, Aventis, etc.) and foreign corporations (Monsanto, etc.) have invested in subsidiaries created *ex nihilo* and specialised in biotechnology, in order to set up in France (e.g. BioSepra, Bachem Biochimie or Diagnostica Stago) or to isolate their biotechnology activity from their core business (e.g. Syral, Biosem or Limagrain Genetics). These firms, which benefit from the captive market of the parent company and the networks and markets to which it affords access, see their turnover increase faster than that of independent firms. Companies in this category are much bigger (in terms of number of employees) and have a significantly higher turnover than those in the other categories. This growth is based above all on the parent company's internal network. On the other hand, the fact that they patent less may not be significant, for some patents may be registered by the parent company.

Group R: New Biotech Firms (DBFs)

This group comprises firms that had been in existence for less than two years at the time of the survey. They are thus constantly evolving. Significantly smaller (Table 6) than firms in the other groups, in terms of number of employees and turnover (often two or three persons, some of whom work part time), DBFs have registered few patents.[9] On the other hand, they have persuaded capital investors to become shareholders (Table 5). Capital investors are often firms in the seed business and are often regional and multi-sectoral. Their capital investments enable biotech firms to survive, for many of these DBFs generate no turnover. They are very seldom engaged in the production of goods for final consumers. By contrast, they have partnerships with research organisations, especially INSERM. The youngest firms are oriented essentially towards the pharmaceutical sector. They are less present in the agriculture, agri-food and cosmetic sectors.

COMPETING BUSINESS MODELS

Three Logics for Three Business Models

The statistic analysis reveals four types of firm which were designed differently from the outset. Firms in Group A base their development on their capacity to produce and transfer their scientific results. For them, a critical resource is access to scientific competencies and techniques developed by academic research. Their development logic can be summarised as follows:

> Scientific discovery that can be transferred--> registration of patent --> entry of venture capital into shareholding to finance an R&D activity--> strong ties with the academic world--> partnerships with pharmaceutical firms to gain access to the market and transfer research--> entry into the new market.

For these firms, often created by researchers from large groups or by academic researchers, insertion in the scientific network is a condition for growth. It is not sufficient, however, for firms must not only develop high-tech research but also transfer and commercialise their results. This often involves research or development contracts with a big company, in which the SME undertakes to provide its partner with specific materials, technologies, know-how or expertise. Relations are formed on the basis of a specific competency recognised by the big firm. The SME's technological lead depends on the quality of its research, and the launching of the activity relies on capital input for the development of the product or process.

Firms in Group B are based on a different logic. Created by researchers or engineers who have identified a commercial opportunity, they focus on specific niches and generate turnover very soon after creation. These firms are part of a local economic fabric resulting from contracts developed by the entrepreneur during his or her previous activity.

The development logic can be outlined as follows:

> Identification of a niche or idea to transfer research --> creation of a firm based on family capital --> commercialisation of products or services to generate a cash flow --> development of business by expansion of the market.

In this logic, the firm's key resources are its first customers and its regional or sectoral position which enable it to be known. Firms in Group C rely on the group to develop their business. They progressively expand their clientele.

Competition or Complementarity

Competition between firms occurs when they are present in the same market. In that case, it is the firms' products or services that compete. Given the emergent character of the biotechnology sector and the innovative nature of the services or products offered, markets are constructed along with supply. Thus, it seems that, at this stage, firms are rivals in their market segment, within the same business model. The market, as usually described, represents a key resource in the development of firms only for those belonging to Group B. On the other hand, all firms are rivals for the acquisition and integration of key resources for development.

Table 7 describes the different areas of competition. As they are based on research and most of them are financing their activities thanks to capital investments, firms on the different business models are not only competing on the markets. They are also competing to convince shareholders (and venture capitalists) to invest in their activities, to develop co-operation with established and promising academic teams (having a star scientist as a Nobel prize as a collaboration or in the scientific advisory board can boost your activity and your share) and to contract with large firms to finance the development and the commercialisation of product.

Table 7: Nature of competition for the different key resources

	Group A	Group B	Group C
Market and cash flow generated by the turnover	International competition within this group. Competition with other actors such as large firms or academic laboratories	Local competition at first, which spreads as the market expands.	These firms compete with certain firms in Group B.
Financial resources and shareholders.	Competition between firms in this group to persuade investors and ***set up a round table of venture capital.*** Competition may also continue when the firm is listed on the stock market.	When investors are in the family there is no competition. However, when the firms apply for local funding and investment from local venture capital firms, there may be competition.	Firms in this group are financed by sales of products and services. Shareholding companies also provide capital.
Relations with the academic community.	Competition between firms in this group to develop relations with academic teams and to sign exclusive licence agreements with them if the university team has patented a technology.	Relations with the academic community are formed on a basis of trust and partnership. Technologies are rarely patented. Niches are narrow and the number of competitors small.	Close relations with the group's research centre.
Relations with large companies.	Competition to sign cooperative agreements.	Few relations with large companies, except as suppliers.	Few relations with companies that are shareholders' rivals.

Resources needed by firms to grow are specific to each type of business model. Thus, firms compete within a business model. But competition between business models remains marginal, both in markets and for the acquisition of resources which, even if they are similar, are owned by different actors.

CONCLUSION

Based on an analysis of the business model of SMEs dedicated to biotech, this paper leads to three conclusions :

• As the biotech sector is still an emergent one, the conditions of the competition between firms differ to the other sectors. Rather than being based of product on the market, firms are in competition for attracting venture capitalists or other sources of capital to fund their development. They are also in competition for contracting research agreements with universities and large firms. But this form of competition is mainly limited to the firms which belong to the business model A. in the business model B, firms look like firms of the mature sector. They are selling products or services and their growth is limited.

• The competition between firms which belong to different business models remains marginal. And it seems that the firm belongs to a specific business model since the very beginning of its creation. Thus, the changing of business model (mainly form B to A) is not easy and it requires a real transformation of the firm in terms of business plan, business project and business actors (collaboration, shareholders, and so on).

• The economic tools of public policy need to be adapted to each business model. For the benefit of the whole economy in terms of job and of wealth creation, each firm of each business model has a positive impact. If firms of business model A need specific tools for biotech DBFs to be designed (research programs, encouraging venture capital, dedicated stock market just as Nasdaq, etc.) the general economic tools for encouraging firms creation (incubators, technological platforms, regional networks, tax reduction for young firms, etc.) will help firms which belong to the business model B. These different tools can be used either at the national or federal for the business model A and to some extend for business model B. The regional economic policy will mainly encourage business model B firms.

Annexes

Characteristic of the Survey

At the initiative of the technology division (Biotechnology group) of the MENRT,[10] a survey was conducted on firms engaged in biotechnology research or development. A questionnaire was sent to 450 private organisations, irrespective of their size or status (listed or not). 221 answers were received but they were largely incomplete. In order to obtain a representative sample, the data base thus obtained was matched with various other available data bases: the base created by the INRA/SERD team, The France Biotech base, the Genetic Engineering Directory, Infogreffe and Diane. Missing information was thus obtained and certain firms were added to the base when all necessary information was available.

In order to standardise the answers, only organisations listed in the trade and commercial register as companies (SA, SARL, SNC) were selected. To guarantee the relevance of comparisons with information published by Ernst and Young (1999 Report), our analysis took into account only those firms with under 500 employees. Lastly, we excluded the rare biotechnology firms founded before 1960. The analysis was finally based on 186 enterprises on which full information was available. The sample analysed can be considered representative, essentially because over 60% of the firms listed in the France Biotech 1998 directory and 90% of the firms in the Genetic Engineering Directory answered the questionnaire. By convention, we consider that the base analysed corresponds to roughly half of the biotechnology enterprises active in France.

The survey consisted of several steps:

1. Compilation of a list of enterprises engaged in biotechnology research, based on available sources;
2. Validation of this list compiled by regional technology representatives, and addition of complementary data;
3. Definition of a list of 450 target enterprises;
4. Administration of the questionnaire;
5. Processing of data and compilation of the directory.

The survey gathered information on several dimensions:

1. Identification of the firm;
2. Ownership and creators;
3. Targeted markets and technologies in use;
4. Patents and certification;
5. Financial information;
6. Partnerships.

List of variables

Creation 1980-90: Creation of the firm between 1980 and 1990
A: Business model "Firms with high growth potential"
sh vent. K : Venture capital firm a shareholder in the SME
sh pers: Shareholder a natural person
sh firm : Firm a shareholder of the SME
Agri/envt: Targeted sector of the firm is agriculture, agro-food or environment
After 90: Creation of the firm after 1990
Before 80: Creation of the firm before 1980
B: Business model: Firms which develop in niches
Patent: hold patents
C: business model: Firms attached to a group
TO 1,5 and 15 Meuros: Turnover between 1.5 and 15 ME per year
TO less than 1,5 Meuros: Turnover less than 1.5 ME per year
TO up to 15 Meuros: Turnover up to 15 ME
CNRS: Partnership with an academic organisation
co-op labs: Academic partnerships
final cons. : Firm supplies goods and services to the final consumer
cosm ou veto: The targeted sector of the firm is veterinary products or cosmetic products
Number of employees between 10 and 50: Number of employees between 10 and 50
Number of employees less than 10: Number of employees less than 10
Number of employees up to 50: Number of employees up to 50
INRA: Partnership with an academic organisation dedicated to agronomic and agro-food research
INSERM: Partnership with an academic organisation dedicated to medical research
Pharma: The targeted sector is human health and pharmaceutical products
Product: The firm sells products rather than services
R: Business model Recently set up firms
Standard: The firm's products are mass-produced rather than customised
All sectors: The firm has no targeted sector. It produces generic research tools used by all sectors
UNI: Partnership with university

Chapter 13

KNOWLEDGE, MARKETS AND BIOTECHNOLOGY[1]

Nico Stehr
VisitingProfessor
Gerhard-Mercator-Universität Duisburg Germany

INTRODUCTION

In this chapter, its argued that the modern economy, as its transforms itself into a knowledge-based economy, loses much of the immunity from societal influences it once enjoyed, at least in advanced societies. This implies that the boundaries of the economy as a social system become more porous and fluid. Among the traffic that increasingly moves across the system-specific boundaries of the economy, from the opposite direction as it were, are cultural practices and beliefs that were heretofore perceived as alien to taken-for-granted conventions of economic conduct and the kinds of preferences immanent within the economic system. The enlargement of the economy is examined with reference to biotechnological products and processes. I will call these changes the "moralization" or "de-commercialization" of the production and consumption process. The moralization of the market and of production ultimately depends on the growing role of knowledge in economic affairs as well as the exceptional rise in affluence and, in its course, consumer sovereignty. Let me explore this line of analysis.

Last spring, Terry Wolf planted half of his central Illinois farm with GM soya seeds and they gave him, he says, an edge in the never-ending battle with weeds. This year, he'll plant one of the hi-tech beans. As Wolf and other farmers across the state prepare for spring planting, many are turning away from GM for the first time since the crops stormed the market in 1995.[2]

The belief that the economic significance of biotechnology exceeds that of the Internet, as Lester Thurow for example maintains,[3] is just as good and as fragile a prediction as is the growing fear of the unknown consequences of biotechnological products. As the title of my paper indicates,

I want to address the interaction between knowledge and the marketplace in modern economies, using the case of biotechnological products as my empirical referent. More specifically, the referent is (new) biotechnological food products (GM foods)[4] already in the marketplace and typically competing with products using different production processes, which the consumer indeed envisions and comprehends as substitute products.[5]

The focus I have chosen also indicates that I do not, in this context, care to discuss the ethics, the politics, the legal ramifications or the need to enlighten the consumer more aggressively about what are seen by some as small[6] and by others as dangerous risks in relation to the potential gains of deploying biotech products and processes in the marketplace.[7]

The point I want to advance is that the current public controversies about genetic modifications of foods are only a small but powerful pointer to more significant transformations of the modern market economy, in which incremental knowledge that is basically contested, conjectural knowledge becomes the motor of the dynamics of the modern economy (see Stehr, 2000). This in turn allows for the possibility that the market is no longer merely a place where preferences are restricted to what is seen as the hallmark of pure "economic" reasoning, that is, a market in which all goods and services are priced according to their utility; and decisions taken by all market participants, and by institutions that modify market behavior,[8] operate according to the same code.[9]

The assertion that the market increasingly allows for what might be called a "moralization" of economic decisions or, to put it differently, that steering functions of the market are no longer limited to purely economic reasoning,[10] runs counter to what many see as the iron logic of modernity: namely the functional differentiation of society, and the inescapable limits, as well as the collective (society-wide) power, of system-specific codes.[11]

What I will describe here as the moralization of the market – without moralizing about an invasion of the market by moral considerations – or a de-commercialization of prices, I see primarily as an outcome of the dynamics of the economic system itself; and not, as some might suspect, of the successful intervention of social movements, the political and the legal system or other societal forces – such as enlightened discourse about the profound paradoxes of modernity – that attempt, as they undeniably do, to impose their logic and their regulatory efforts on the marketplace. The moralization of market conduct is more than merely the presence of greater and lesser uncertainties or risks associated with human conduct. In addition, I want to sketch the societal processes underway in their broad outlines.[12] I do not attend to the complex details and contexts of different biotechnology applications, controversies and prospects in different countries.

The perspective I want to advance requires that I first describe the logic of modernity as it manifests itself in the theoretical perspective of (societal) functional differentiation, that is the centrifugal tendency inherent in all modern societies. A second, and competing, view relevant in this context suggests the opposite: namely, that it is economic rationality that conquers and controls society. Although these views of the modern economy are diametrically opposed, they both testify to the inherent power and authority of economic reasoning. In the case of differentiation theory, the social system of the economy – as first articulated as a political demand by the proponents of mercantilism – ought to be and in fact is self-sufficient: it administers its own affairs and does so on the basis of a logic or code that is uniquely its own. It follows that new developments within the economy must also submit to the same logic. In the case of the idea that we are faced not by a differentiation of social codes, but rather by an economization of society, it follows too that new biotechnological developments surrender to the logic of the market.

THE CIVILIZATION OF CAPITALISM[13]

"In capitalist reality as distinguished from its textbook picture, it is not (price) competition which counts but the competition from the new commodity, the new technology, the new source of supply, the new type of organization." (Schumpeter, [1942] 1962:84)

In response to the elementary question "What is the economy?" or even "What constitutes the modern economy?" there are two key features of economic conduct that are almost always mentioned as constitutive of economic conduct, and that also happen to be of particular interest in this context. Citing Emil Lederer (1922:18): nothing, so it seems, is more straightforward than to enumerate the essential features of the economy.

It is, first, quite evident to us that the economic system differs from other social institutions even though one is also forced to attend to economic issues in churches, in families, in city councils or in universities. Economic activities as a distinct form of social conduct, in addition, satisfy (material) human needs; or economics "examines that part of individual and social action which is most closely connected with the attainment and with the use of the material requisites of well-being" (Marshall, 1920: 1; also Sombart, [1916] 1921: 13).

Second, the governing professional images of economic activities imply that economic conduct and market institutions closely follow the rationality principle, generating the optimal gain with the smallest possible input. Since such profits/benefits ratios are not generated in isolation, the kind of social interactions and communications that typify "genuine" economic activities are those encountered within the boundaries of internal economic markets,[14] where rational decisions and self-interested behavior are put into practice.

What is missing from these accounts is the fact that the economy is a dynamic system, and that the changes are accelerating. Moreover, the boundaries of the economic system may be much more porous than is implied in this account of the economy as a functional subsystem with its own unique logic and media of communication.[15] In particular, the motives, values or preferences of at least of some of the salient actors within the economic system may be much less restrictive than is implied in the predominant account (cf. Douglas and Isherwood, 1979:56-70). In other words, using the Marxists' imagery, the nature of the cultural complement of the modern economy is perhaps changing as rapidly as its substructure. And the missing dynamic attributes may be related, as I will argue.

The changes I want to consider may not only be linked, but may also lead to a reversal of the commonly assumed relationship between substructure and superstructure. The much more complex or varied superstructure is not merely dependent or a "cultural complement", as Schumpeter ([1942] 1962:121) describes the derivative culture of capitalism; it is, rather, the motor of the transformation and the reason for the trajectory of the dynamism of capitalism.

THE LOGIC OF MODERNITY

The emergence of the economic order as a distinct social system, in the first half of the nineteenth century, widely hailed as a moral accomplishment itself, is one of the hallmarks of modernity.[16] The economy was seen and comprehended in terms of its own laws and causalities. Political interventions should be limited to, and should respect as well as nurture, the unique contingencies of economic action. In contrast, the moral domain, at least in those days, was seen as the proper sphere of intervention, programmes and regulation by politicians and others, all designed to mold the character and conduct of moral subjects (cf. Rose 101-107). What is noteworthy about these developments, aside from the particular understanding of the moral domain, is the strict separation of these spheres of conduct.

The most radical contemporary proponent of social differentiation as the logic of modernity is Niklas Luhmann. The perspective of functional differentiation favors dichotomies and exclusive logics of social action in differentiated social systems. Luhmann defines society as a system that operates on meaning, and therefore as a communicating system that is integrated on the basis of meaningful communication. Society reproduces itself on the basis of communication. Its unity is the autopoiesis of communication. In contrast to the societal system, subsystems also communicate with other subsystems that form part of their environment. The identity of subsystems is constituted on the basis of a differentiated principle or special code. In the case of the economy, according to Luhmann, this happens to be the principle of payment (and the opposite of the same code, namely non-payment). Payments have exactly all the characteristics of autopoietic processes: "They are only possible on the basis of payments and in the context of the recursive economic context of autopoiesis they have no other meaning but to assure payments" (Luhmann, 1988:52). Payments (in the form of money and prices) reproduce the economy. These processes assure that the economic system is both a closed and an open system (in the sense of a future direction). The result of these reflections is to transform all economic categories that otherwise constitute core categories of economic discourse into derivatives of payment processes. However, this also implies that many elements that are usually considered to be part of the economic system are not part of Luhmann's discourse about the economy. This applies to resources, that is, to commodities and services for which payments are made, or to the psychological dispositions of actors, for example. These features or processes are part of the environment of the economic system. The economic subsystem, however, is the location in which communication about the elements takes place.

The realities of the economy are dramatically changing, and with them the classical as well as the radically modern perception of the economy. It is not only that national economies are inhabiting a much wider domain or space, commonly discussed under the heading of globalization, but also that the age of the "material" economy is giving way to the "symbolic" economy driven by the "immaterial" resource or "raw material," knowledge.

The Knowledge-Based Modern Economy

To an increasing degree knowledge, rather than labor or property, is constitutive for economic and social activities. Knowledge becomes the source of the possibility of economic growth and competitive advantage among firms and among entire societies and regions of the world. The developments in question allow us to speak of the transformation of modern industrial society into a knowledge society. Even more generally, the knowledge society represents a social and economic world in which more and more things are "made" to happen, rather than a social reality in which things are simply "happened".

Table 1. The Knowledge-Based Economy, 1995/1996

	Investments in knowledge [a] compared to (physical investment) as % of 1995 GDP		Value-added of knowledge-based industries [b] as a % of the 1995/1996 business sector value-added
Italy	6,1	(18,0)	41,3
Japan	6,6	(28,5)	53,0
Australia	6,8	(22,6)	48,0
Germany [c]	7,1	(21,4)	58,6
OECD	7,9	(20,1)	50,9
EU	8,0	(19,0)	48,4
USA	8,4	(16,9)	55,3
UK	8,5	(16,3)	51,5
France	10,2	(17,9)	50,0
Sweden	10,6	(14,6)	50,7
Canada	8,8	(16,9)	51,0

[a] Total investment in knowledge is calculated as the sum of R&D expenditures (minus equipment), public spending on education, and investment in software (minus household purchases of packaged software).

[b] The OECD includes all firms that are relatively intensive in their inputs of technology and/or human capital. From the service sector, communication firms, finance, insurance and business services, as well as "community, social and personal services" are included.

[3] West Germany.

Source: OECD (1999: 114-115)

It is therefore necessary to ask whether the economic principles and policies developed, tested, and adopted for the realities of the material economy are applicable to the new economic realities. In other words, the conditions that allow for the economic transformations under consideration also render traditional economic discourse (and policy derived from such premises) about economic affairs less practical and powerful. In particular, one has to ask whether the claims, regularities and principles now encountered in economic discourse have a bearing on the dynamics of the fabrication, distribution and consumption of knowledge in economic processes.

I will point to but one salient attribute of knowledge that is of particular importance in this context:

KNOWLEDGE ABOUT KNOWLEDGE

I would like to define knowledge as a capacity for social action and as a model for reality. In this sense, knowledge is a universal phenomenon, or an anthropological constant. My choice of terms derives from Francis Bacon's famous observation "scientia est potentia," or as it has often been translated in a somewhat misleading fashion: Knowledge is Power. Bacon suggests that knowledge derives its utility from its capacity to set something in motion. The term potentia or capacity is employed to describe the power of knowing. Knowledge assumes significance under conditions where social action is, for whatever reasons, based on a certain degree of freedom in the courses of action that can be chosen.

The extraordinary importance of scientific and technical knowledge does not primarily derive from its peculiar cultural image as representing essentially uncontested (or, objective, that is, reality-congruent) knowledge claims, in the face of which many segments of the public are ready to suspend doubt about its fragility. In this context, the tremendous importance of scientific and technical knowledge in developed societies is related to one unique attribute of such knowledge, namely that it represents incremental capacities for social and economic action, or an increase in the ability of "how-to-do-it" that may be "privately appropriated," if only temporarily, since the benefits from innovations based on incremental knowledge and new knowledge itself are extended or leak to third parties.

But unlike the conviction displayed by classical accounts of scientific knowledge, in many instances science is incapable of offering cognitive certainty. Science cannot offer definitive or even true statements (in the sense of proven causal chains) for practical purposes, but only more or less plausible and often contested assumptions, scenarios, and probabilities. Instead of being the source of reliable trustworthy knowledge, science thus becomes a source of uncertainty. And contrary to what rational scientific theories suggest, this problem cannot be comprehended or remedied by differentiating between "good" or "bad" science (or between pseudo-science and correct, i.e. proper, science). After all, who would be capable of doing this under conditions of uncertainty? Knowledge is essentially contested and the standard areas of contestation around knowledge-based products in general and biotechnologies in particular are those of expertise, risk, responsibility and regulation (cf. Gofton and Haimes, 1999:2.7).[17]

Wealth and Consumer Sovereignty

Despite the recurrent concerns about the performance of the economy, I would like to point out that nothing in the history of the industrialized countries in Western Europe and North America resembles their experiences between 1950 and 1985. As Alan Milward (1992:21) has stated:

> [b]y the end of this period the perpetual possibility of serious economic hardship which had earlier always hovered over the lives of three-quarters of the population now menaced only about one fifth of it. Although absolute poverty still existed in even the richest countries, the material standard of living for most people improved almost without interruption and often very rapidly for thirty-five years. Above all else, these are the marks of the uniqueness of the experience.

The emancipation from economic vulnerability and subjugation of large segments of the population, which Marx and Engels had not foreseen and which does not occur to the same extent and at a similar pace in all industrialized countries, provides for the material foundation of new forms of inequality (see Stehr, 1999). More concretely, what diminishes is the tightness of the linkage in the material dependence of many actors on their occupational status alone; and what increases is the relative material emancipation from the labor market, in the form of personal and household wealth. The growth of individualism in modern society provides the cultural counterpart for the rise of affluence.[18]

Economic discourse today continues to be linked to the eighteenth-century definition of the major factors of production; namely, capital and labor, with their combinations and their consequences measured in monetary units. And to this day, in much of social science, consumption phenomena and practices tend to be seen as derivative of the core institutions of society, such as social class, the state and culture, which are in turn shaped by production and employment.

However, if the focus on, and the implied equivalence and equilibrium between, production and consumption shift from work (in the narrow sense of the term) to forms of life of employees and households in modern society, then an analysis of the consumption side – especially in relation to total wealth and life expectancy – is more pertinent than is the mere income of individuals or households.

What a society makes still matters. But how it consumes what it makes increasingly matters more.[19] The growing degree of consumer sovereignty also affects the social organization of work and the ways in which products are generated through greater interactive contacts with consumers and clients. At least it changes the world of work for those who have to deal directly with the heightened self-confidence of the consumer in the marketplace (Frenkel, Korczynski, Shire and Tam, 1999:66-81) and for those charged with realizing and delivering inventions.

BIOTECHNOLOGICAL PROCESSES AND PRODUCTS

The products we buy today are often pretty much the same as the ones that hung in our closets or found their way onto our tables a decade ago. However, the ways in which these products and new goods are produced have been radically altered. Biotechnological developments, and agricultural biotechnology in particular as an object of publicly and commercially funded research now almost two decades old, are paradigmatic of the knowledge-intensive economy. It is true that in some countries, there are still modest pockets of non-biotechnology; on the whole, however, according to a disputed claim by Buttel (1999:1.2, emphasis added; cf. also Buttel, 2000)[20] in "mainstream agricultural research circles across the globe biotechnology (or 'genetic engineering') is largely the accepted approach". However, success in research and success in moving its products to the market, making them acceptable and mastering the market represent very different stories.

In discussing the moralization of the market in the case of biotechnology it is useful to distinguish between process (generating products) and product, even though one of the characteristic features of the knowledge-intensive economy is the fusion of process and product. The moralization of the market occurs from the bottom up. It occurs in response to the products that appear at the market and focuses on the processes on the basis of which the products are generated.

The process is knowledge-based. But the inherent uncertainty and endemic conflicts that surround knowledge systems, the lack of confidence or trust that extends to expert opinions, the freedom of choice without guarantees and the growing reflexive knowledgeability of consumers all make it highly likely that market decisions will not only be governed by risk aversion but also by moral and ethical considerations.

Efforts that assist consumers in making "informed choices" are generally considered and advocated by regulators as the most rational (promising) approach for consumers in making market choices.[21] Under certain circumstances, information and education will not work.[22] The unanticipated consequences of programs advocating informed choice could well multiply uncertainty about the biotechnological products and processes. Consumer studies indicate that consumers respond to and resolve uncertainties, lack of confidence, and the complexity and volume of information in various ways, involving references to a complex set of values and social networks.[23]

CONCLUSION

What I have described as the moralization of the consumption of products, it is of some historical interest to note if only for the purpose of contrast, represents an era in the evolution of the economy far removed from the 1920s notion of "mass consumption" that could be engineered according to the principles of a kind of social Taylorism. What was needed, it was thought, for example, by Alfred P. Sloan, was the scientific management of the needs, desires and fantasies of the consumer (see Webster and Robins, 1989:334-336). One outcome of these ideas was the immense growth of market research, advertising and marketing. It is doubtful that traditional advertising and marketing tools will be very effective in an era of growing consumer independence and knowledgeability.

The moralization of markets for products, or the de-commercialization of prices, does not imply that markets are collapsing, that competition is less efficient, that prices can no longer fulfill the function they perform in the economy, or that the freedom of consumers or producers is seriously undermined. But what it means is that, as a result of fundamental transformations in the economic system and society, the modern consumer's choice of products in the marketplace increasingly reflects considerations other than those of bare "utility". We have seen consumers doing so at a growing pace; preferences expressed by choosing "fair" and "organic" products were but the beginning.

Chapter 14

BIOTECHNOLOGY AND POLICY IN AN INNOVATION SYSTEM:

Strategy, Stewardship and Sector Promotion

John de la Mothe
PRIME, University of Ottawa

INTRODUCTION

On March 14th 2000, U.S. President Bill Clinton and British Prime Minister Tony Blair jointly announced their commitment to making all raw genetic data free. The next day the stock markets reacted to this apparent threat to intellectual property rights and patent protection by putting corporate biotech stock prices into a nosedive, losing 20 per cent or more, and The Financial Times editorial chastened the pair for taking a blatantly populist position for political gain. (The Financial Times, March 16 2000, p.10) These reactions to a single announcement point to what many believe to be the core issue for biotechnology policy – the ability of governments to navigate the treacherous waters between economic growth and ethics.

As many chapters in this book have already demonstrated, the economic dynamics and impacts of the myriad activities collectively referred to as 'biotechnology' portend to offer considerable (if not vast) employment and commercial benefits. It is thus not surprising that many pundits in the private sector wax enthusiastically about the potential of this new 'frontier' of research if market forces are allowed to fuel investments and competition, unbridled – to the extent possible - by government hands. Yet, the nature of the technology – e.g. potentially changing conceptions of ourselves, permitting the development of techniques which no one knows with certainty what the ecological effects might be, etc. – have led to legitimate public debates.

Central to these are deeper concerns regarding ownership between biotechnology as a public good or as a private asset. Debates over ownership typically pivot on questions about property rights, public access, and the equitable distribution of benefits derived from advances in bio-techniques, such as gene therapy. Of course, prevailing legal conceptions of property and notions of ownership include the rights of possession, use, management, income, security, transmissibility and exclusivity. (Gould 1993, p.172) In other words, a public fear is that governments will lose control of this transformative and soon-to-be pervasive technology.

Ethical concerns with biotechnology clearly cover a broad swath. If one could try to typify these, a spectrum might range from moral issues (in which case cloning and other biotechniques is felt to put 'Man' in the place of God), to basic anti-technology sentiments (as in the media representation of 'Frankenfoods'), to a fear of the unknown – or, more precisely, to an sense that biotechnologies will increase uncertainty and thus, public risk, to a core anti-capitalist or anti-corporatist reaction – particularly when it is felt – or shown – either that there is 'excessive' (to some minds) profit taking in certain biotechnology sectors or that the benefits derived from biotechnologies in terms of nutrition, agriculture, and health are not only not being distributed to the developing world but are actually having negative impacts on these impoverished regions. Concern with risk management is probably the most legitimate of these.

But these together merely point to the dynamic interface between policy, business and society with this evolving technology. They certainly point to the environment within which policy makers must work. But they do not usefully illuminate the operational challenges of policy making. With biotechnologies affecting so many sectors of the economy, and with so many government departments dealing with aspects of biotechnology, the multidimensionality and complexity of governing can only be assessed contextually (i.e. within the larger context of innovation). Seen thus, the goals of policies for biotechnology would be set strategically, directed towards a twin commitment to stewardship and sectoral promotion. But how can this dynamic, contextual visualization be set out?

BIOTECHNOLOGY IN THE CONTEXT OF AN INNOVATION SYSTEM

A useful step is to recognize that biotechnology is often considered largely in isolation from the specific national contexts that, to a large extent, determine successes or failures in taking advantage of what the new technologies have to offer. The intrinsic capacity of a country to stimulate technological change and innovation – and hence to integrate biotechnology into those processes – has rarely been taken into account in formulating biotechnology policies and strategies. In other words, in many countries policies, publics and vested interested have been reactionary, not integrative. This need not be the case.

The Innovation System (IS) approach provides a useful framework for elucidating the national context in which biotechnology should be integrated. It is also useful in highlighting the complexities of the process of knowledge production, technological innovation, and diffusion.

Figure 1 provides a simplified framework for understanding the knowledge and technology flows based on the idea of an innovation system. The framework encompasses a network of units, systems and sub-systems which interact to generate, exchange and distribute knowledge. The effective functioning of the system depends in part on the capabilities and characteristics of its individual units. It also depends on the nature, frequency and intensity of the linkages and flows of biotechnological knowledge, technology and information among the different units and sub-systems within the system. For our purposes, the terms linkages or flows are synonymous with that of technology or technology transfer.

In Figure 1, research, technology development and diffusion are linked through the research, production and distribution systems. Biotechnology research may be basic, applied or adaptive and should be closely integrated with national science and technology objectives. Close interaction is necessary between the research and the respective communities of practice, both in order to identify the major production or problem areas to which research should be given priority, and in providing feedback on the acceptability or appropriateness of biotechnology products and processes that are generated by the research and industrial communities.

Figure 1.

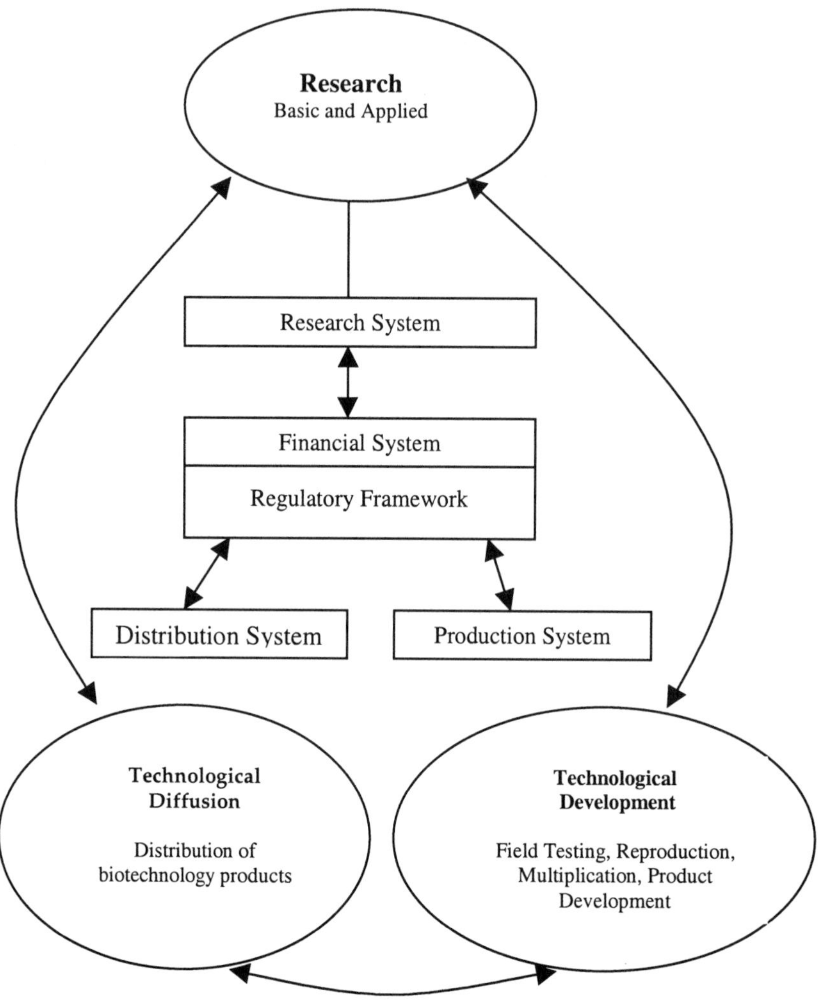

Development encompasses the activities which translate the results of successful laboratory research into a tangible biotechnological product, such as a genetically modified seed or disease-free planting material. These may include small and large-scale field testing, seeds multiplication, or setting up a pilot plant. Product development in biotechnology can variously involve both public and private actors: commodity boards, farmers, firms, and so on. Again, as with research, interaction and feedback is an important part of biotech development, and of the overall innovation system.

Technology transfer may take place – both as commercial and non-commercial transactions – in all phases of the knowledge production, technology development and diffusion process suggested in Figure 1. In biotechnology, the forms of technology transfer encompass: education and training; the acquisition of research techniques, material and equipment; the acquisition of biotechnology products (such as GMOs or plant varieties), etc. In an effective innovation system, feedback would occur between the final users and the research system, and between the final users and the production system.

Many different public and private actors and institutions interact between the research phase, the product development phase, and the diffusion of new technologies and techniques to final or intermediate users. In the policy realm, these include government departments responsible for industry, health, agriculture, natural resources, scientific research, national research councils, regulation and intellectual property, universities and colleges, as well as provincial or state equivalents. Interaction with the public, NGOs and with industry, as well as international competition and regulatory regimes fill out the biotechnology innovation system (IS) and thus, as will be sketched below, calls for coordination, transparency and participation from a policy perspective in order to achieve an effective IS.

In contrast with mechanical and chemical technologies which have been developed by private firms in industrialized economies, biological techniques have a stronger 'public good' element. As chapters in this volume have already noted, biotechnologies rely heavily on the research (R&D) or knowledge production system found in universities. This knowledge intensity and interaction is an important feature.

The different units, systems and sub-systems shown in Figure 1 function within the confines of particular environments – policy, financial and regulatory. Elements of that environment which are of particular importance for biotechnology include: macro-economic policies (especially strategic structural adjustment, liberalization and harmonization) and their link to the micro-environment; levels of investment and incentives; science and technology policies; specific sectoral policy areas such as agriculture, environmental and natural resource policies; regulatory and IPR regimes; and so on.

An important feature of emerging biotechnology innovation systems will increasingly be their degree of openness. Elements of knowledge, technology and information may be acquired from a diversity of sources, at several different levels. Interaction and feedback in the research, technology development and diffusion processes thus occurs not only at the micro-level, between units forming part of the system, or at a national level, but also at regional and international levels.

In the above sketch, the context for biotechnology was cast in terms of a number of key elements. Knowledge, technology and information were here seen as the core currency. Flows and transfers of these were seen as the avenues for value creation. Capacity in terms of research, production and distribution was an important parameter for success or effectiveness. The environment – which could stimulate or stifle – presents an important policy realm. And the zest of interaction and the fit of linkages between actors was seen as the essential measure of health for the innovation system, keeping in mind that sector specificities – in this case, for biotechnology – key to analytic and operational refinement. With this framework in hand, we might be allowed to say that we now have a solid schematic or diagnostic tool through which to assess the status of a country's biotechnology innovation system. But what utility does it hold, can it be used, and what can be said about developing a strategic policy framework for biotechnology?

BIOTECHNOLOGY AND POLICY IN AN INNOVATION SYSTEM

The utility of the innovation system framework, from a policy perspective, can be suggested in the following way. By emphasizing innovation, and not simply science or technology, one immediately moves into a dynamic – not a static – mode. By focusing on the interactions between actors, one breaks away from a reliance on input-output assessments and moves instead towards a consideration of flows and processes. This alone shifts the attention of policies and public programs towards the performance of networks and clusters, and towards the fostering of partnerships (as in pre-competitive research consortia, networks of centers of excellence, and so on). Put another way, in a static policy framework, one is inevitably drawn towards unrealistic linear models which try to reveal the complexities of the technology-economy interface in a two dimensional way. It typically attends to how much R&D spending is taking place is a gross level rather than looking for the synergies, loci of creativity and entrepreneurial energies, etc. The institutions are targeted instead of the exchanges. In a static framework, macro-indicators of performance – such as the GERD/GDP – allows for little more than a sloppy comparison at the national level. But by dealing with an IS view, governments become the facilitators - they back leaders instead of picking winners. Sectoral competence blocks become highly visible, thus allowing supportive and niche or customized policy actions. In the IS perspective, firms are cast in their proper central role as creators of wealth and R&D is portrayed as an important but not the only factor. In other words, by using an IS framework one can quickly move towards a different type of industrial policy or political economy. These general shifts are highly useful and a variety of policy actions – such as putting computers in schools, connecting communities, extending patent protection, and so on – can all be powerfully comprehended within an IS framework. This could not be said of a traditional industrial policy framework. In sum, an IS is both analytically and proscriptively of value.

But having said this, how can it be used in as a policy tool? Unlike firms – which, even in multi-technology/multi-product organizations, are fairly single-minded in their strategy (to compete against other firms) – governments and their policies are continually operating under multiple, sometimes competing, commitments. For example, in the area of biotechnology, and as stated at the outset, this requires a twin commitment to stewardship and sector/technology promotion.

Seen in more concrete terms, governments carry out and support science and technology activities in order to achieve the following.

- provide technical assistance to small and medium sized businesses which are working in a technology-intensive area and which do not have the needed in-house expertise or equipment. This is an important role for governments which enables firms to grow, compete, and in turn create new value-added jobs. No firm or university can easily provide this service.

- pursue new technology development in areas such as rDNA

- establish and negotiate standards in order to harmonize domestic environments and international regimes in order to provide a favorable business climate. Again, state-to-state negotiations cannot be done by firms and government science in the public interest is needed to ensure level playing fields and to avoid conflicts of interest.

- undertake testing and approval in areas related to drugs, bio-medical devices, vaccines, blood products, and the like which clearly require government involvement as well as a research capability in order to evaluate and verify outside results for the protection of the public.

- undertake environmental and agricultural monitoring for the protection of Canada's eco-system and commons (in support of existing environmental standards and in anticipation of the identification of new environmental threats). The capacity of the government to carry out such work is critical as ecological threats emerge and as the government commits to meeting negotiated international treaty levels which would be difficult to contract out. Moreover, the capacity to conduct survey work and stock assessments in order to understand changes in the agricultural and ecological systems of a nation (including the fisheries), geological transitions, and so on are key and are germane to government— not industry —goals and mandates.

- support emergency preparedness. Again, firms operating for profit would be hard pressed to undertake earthquake modeling and monitoring over the long haul and the public would rightly wonder if emergency preparedness, operated by the private sector, would provide the responsiveness, warning, and universality that citizens require.

- support policy in the science-based departments and agencies as well as in industry, health, environment, natural resources, and so on. To farm all these responsibilities out to academic or private sector concerns would not only create a government contract monitoring and management nightmare but could also lead to breaches of security, a de-coupling of government science from government policy, and a lost assurance that government and the public interest were matched.

- continue regulatory monitoring and compliance activities such as monitoring and regulatory control of food, drugs, consumer product safety, and the like.

- conduct basic research, not because government researchers should be expected to contribute to the international open literature, but because basic research will support government researchers to be involved in the latest developments, findings and techniques, and will keep vibrant an external research network which can be called upon in support of government science. Active research will serve to promote an attractive career path for researchers in which valuable scientific and technical work can be carried out, thus ensuring the revitalization of government science.

These are but a few concrete examples of the legitimate multiplexing roles for government in science and technology, all of which can be effectively contextualized within an IS.

But brought more directly into the question of biotechnology and policy, as Figure 2 shows, the research base underpinning biotechnology includes such fields as molecular biology, genetics, biochemistry, bioengineering, and immunology. Increasingly we are witnessing too cross-over or 'fusion' technologies such as bioinformatics. This is not surprising of course since advances in gene sequencing deeply require super-computing and the ability to massage huge data sets.

Figure 2.

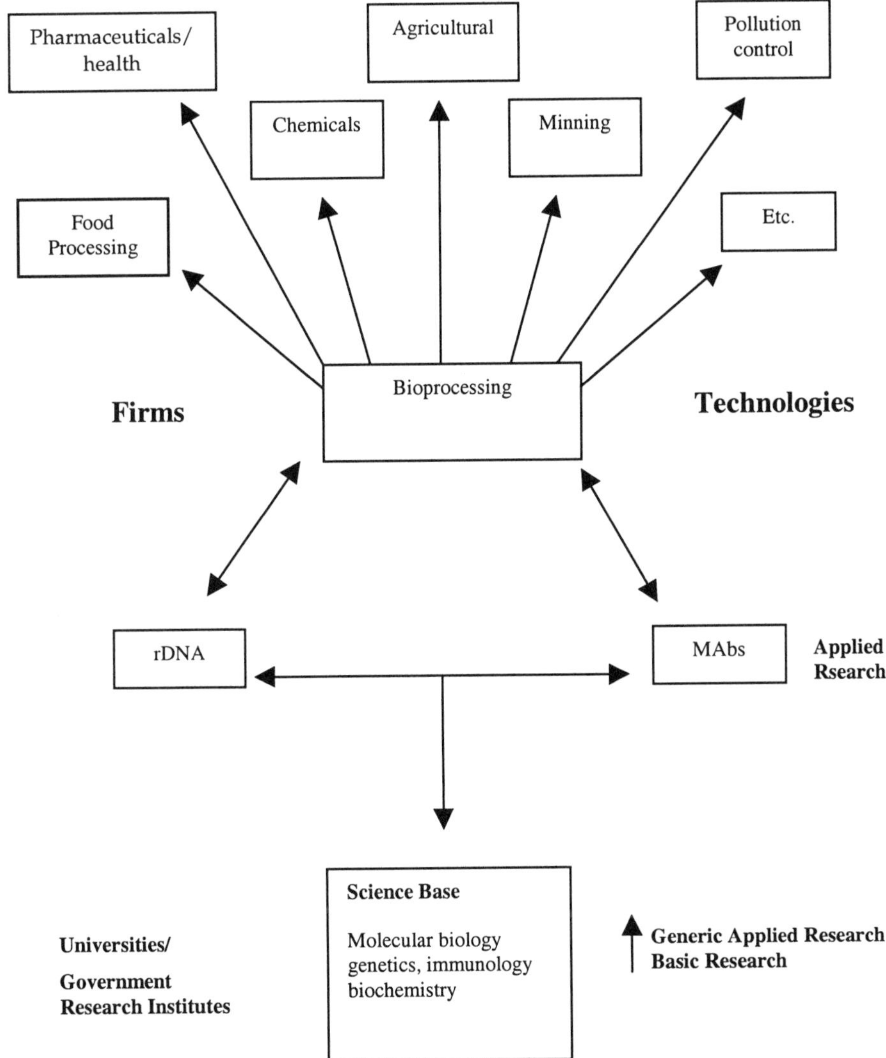

From the research base, which governments must support vigorously, we arrive – rDNA and Mabs – at bioprocessing, a key enabling technique. The impact of bioprocessing can be felt in food processing, pharmaceuticals, chemicals, agriculture, mining, pollution control and several other areas of firm and technology development. Thus, the use of an innovation system framework permits the contemplation of 'mapping' linkages and flows by sector and in so doing allows for the design and pursuit of a strategic policy.

In light of the above, what can be said about establishing a strategic policy for biotechnology?

In answering this query we must ever be vigilant of the fact that public concerns are real. In other words, the goal of government policy should not blindly privilege industrial interests or technologists' dreams but instead must embody a concern with human dignity, a commitment to safeguard bio-diversity, human protection and health, recognition of the individual's right to make informed decisions about his or her use of biotechnology, and an interest in the ethics of testing (as in animal testing).

But more particularly I would highlight the following areas.

- Governments must take public involvement in decision making and setting priorities and directions extremely seriously. This is not simply to promote industry interests but is to – using an IS approach – develop a true partnership in technology development. Many governments exercise consultation as a sham or as window dressing (as allowing interested or concerned groups simply to 'participate in participation'). The benefits of true openness will be manifest.

- Research must be vigorously supported, both through direct investments and through incentives such as leveraged funds, tax benefits, patent and intellectual property protection, and so on. Beneficial advances will come from the base, and research still surprise based.

- The various regulatory regimes must be attended to, sharpened and harmonized. Given the harsher public and regulatory environment in Europe, it is not surprising that there has been a notable flow of German activities to the U.S. If policy makers wish to capture the benefits to downstream research and developments, then regulation is worth a close and integrated second look. This means that priorities need to be set in the national interest.

- In setting priorities a number of common areas emerge across countries – such as genomics, genetic engineering, protein engineering, vaccines, and bio-diagnostics. Policy makers need to take national capabilities into consideration when setting directions.

- Policy relevant data sets and statistics are widely needed by governments in order to map their sectoral IS and address gaps or bolster existing strengths.

- Of course, these areas are not independent of the general needs for human resources, better IPR, or the clear need for responsible leadership in biotechnology.

CONCLUSION:

The broad goal of biotechnology policy is to simultaneously protect the public while promoting the sectors which produce, use and diffuse these techniques. As noted in the opening chapter, the benefits of this realm of technology include improved disease diagnostics, new drug therapies, foods with health benefits, higher yield crops which are pest resistant, renewable fuels, bio-remediation, and so on. Within this chapter, we have argued a number of related points. It was suggested that by focusing on innovation systems (IS) as an analytic and diagnostic framework policy makers can design and deploy niche strategies, programs and policies. It was suggested that there are at least 5 points which are worthy of attention if a strategic biotechnology policy is to be developed in a national context. And finally, a point that I will re-iterate here, it must be remembered that no policy for an area as diverse and dynamic as biotechnology will succeed unless there is a firm policy commitment to horizontal coordination across government agencies, there is true transparency and accessibility to the public, and a clear recognition that such policy making is not a one-shot deal but is instead a continuous process of renewal.

PART V

CONCLUSION

Chapter 15

ISSUES FOR FUTURE RESEARCH, MEASUREMENT AND POLICY

John de la Mothe
PRIME, University of Ottawa

and

Jorge Niosi
CIRANO, University of Quebec at Montreal

The present volume vividly illustrates the many different theoretical and methodological approaches that have been used, in the last twenty years, to study the emergent field of economics, management and sociology of biotechnology. Some perspectives have put forward the institutional support that new, small biotechnology firms require in order to grow. Others have insisted on the constellation of competencies (inside the firms and around them) that are necessary for the successful growth of the new dedicated biotechnology firms (DBFs). In the first part of this conclusion, we recall some of the most frequent theoretical frameworks that have been used in the analysis of biotechnology, and underline the need for integration. The second part of the conclusion summarizes some of the tools that have nourished these studies, as well as some of the shaded methodological areas. Finally, we suggest that biotechnology may follow the same path as most of the earlier knowledge-intensive industries, a path that heads towards increasing economic concentration and the rise of dominant firms.

THEORETICAL ISSUES

In the analysis of the new biotechnology, several theoretical approaches have proved useful: competence blocs, technological systems, network analysis, national systems of innovation, critical success factors, and resource and competence theories of the firm. Table 1 summarizes some of the most widely used conceptual perspectives, and names some authors that have put forward a particular perspective (see Table 1).

Table 1. Comparative theoretical frameworks for the study of biotechnology

Theoretical framework	Authors putting forward the perspective
Competence or development blocs, competencies are embedded in groups of actors; demand is key and drives the system. Competence blocs may be formed by local groups of actors	G. Eliasson (in this volume)
Technological systems as networks of diverse actors; flows of technical knowledge are their substance.	B. Carlsson (in this volume)
Regional clusters structured by knowledge spill-overs among firms and other institutions	J. Niosi and T.G. Bas (2000) M. Prevezer (1998) L. Zucker et al. (1998a) de la Mothe and Paquet (1998)
Resource and competence theory of the firm, Critical success factors. Emphasis is on internal resources and competencies	J. Niosi, V. Mangematin (in this volume)
National systems of innovation	S. Batholomew (1997) M. McKelvey (1996) J. Senker et al.(1996)
Ecological analysis	A. Ryan, J. Freeman & R. Hybels (1995)
Evolutionary Economics	M. McKelvey (1996); L. Orsenigo (1989)

These different perspectives seem to us more complementary than rival. For instance, as Carlsson points out in this volume, high technology industries tend to be regionally concentrated, suggesting that competence blocs are local (Carlsson, p. 24). Thus development bloc theory can be used for regional cluster analysis. Also, competence blocs are networks of agents exploiting economic complementarities for the production of a set of goods and services. Technological systems are also such types of networks, but relationships among actors are here only technological (ibid, p. 11), Technological systems may thus be conceived as a subset of development blocs.

Similarly, national systems of innovation have often (but not always) been conceptualized within the framework of evolutionary economics, an approach that is close to ecological analysis, and usually includes technology systems (Nelson, 1993; Niosi, Saviotti, Bellon and Crow, 1993). On the whole, evolutionary perspectives, as well as different approaches considering a variety of actors, institutions and networks, seemed well positioned for the analysis of this emergent activity. However, the challenge of integrating these partial approaches is still there, as each of them – as in the fable of the blind people and the elephant -- seems to convey some part of the truth.

Methodological Issues

Methodological issues are many and their weight seems extremely high. What firms must be considered dedicated biotechnology firms? In this volume, Arundel appropriately argues that we do not yet have a good definition of biotechnology and that, usually, both old and new biotechnologies are subsumed under the same conceptual umbrella. Also, it may be added that divergences persist even when biotechnology is defined in a more narrow way, putting forward the processes that are characteristic of the "new" biotechnology. Some authors have highlighted "genetic engineering" as its key dimension (McKelvey, 1996) while others have preferred putting "recombinant DNA" at the center of their definition (Kenney, 1986:2).

Do DBFs constitute a new industry? Some authors, such as Eliasson in this volume believe that they do; others, such as Arundel and Rose also in the present book, insist on the fact that biotechnology is a generic technology, used in many different industries. This fact explains the difficulties that we experience in adequately measuring its diffuse economic and social impacts.

What are the best available research methods required exploring the new technology and activity? Table 2 summarizes the methods shown in this volume and in related literature (see Table 2).

Table 2. Methods: Patents, surveys, databases, publication and citation analysis

Methods	Advantages	Disadvantages, dangers	Authors using the method
Patent analysis	It thrives on high propensity to patent by DBFs. Captures location of inventor Tracks flows of knowledge	Different patents have different economic value	J. Niosi and T. G. Bas (2000) R. Oliver (2000)
Surveys based on samples	New knowledge can be produced	Some key information may not be disclosed Sampling errors	Mangematin, Niosi, Yarkin (in this volume)
Publication citation	Key scientific supply can be tracked Networks of innovators can be unveiled	Some key knowledge is tacit	L. Zucker et al. (1998b)
Databases, population surveys	Larger population of firms covered	Population errors	P. Hakansson et al.(1993); Arundel; Rose (in this volume)

It may be safely argued that all these methods have yielded good results in the past, promise to bring similar payoffs in the future, and are probably complementary, rather than concurrent. Studies based on samples may explore new issues for future surveys of entire populations of firms. Patent data may serve to locate clusters, promising technologies and competitive firms. Publication and citation data have served to find out star scientists, those that often produce revolutionary technologies and create successful firms.

However, measurement problems are not confined to these areas. As Arundel points out in his chapter, little is known about application of biotechnology products by firms that do not conduct biotechnology R&D. (But Statistics Canada has done some pioneering work in this area). Also, we have scanty knowledge of alternative technologies that may have been displaced by new biotechnology with the associate loss of employment and revenues. Similarly, the economic and social benefits of biotechnology are still unclear because most R&D projects stretch over many years, with few of them having yielded measurable economic and social returns. In other words, the economic impacts of biotechnology can not yet be estimated using the established methods displayed in Table 2.

WHAT FUTURE FOR BIOTECHNOLOGY FIRMS?

Biotechnology is a science-based activity. All science-based industries and firms operate under increasing return conditions, namely high initial R&D and marketing costs, that need to be spread on the highest possible output, in order to reduce unit costs of products. Thus the drive to continuously increase the firm's production. Other factors supporting increasing returns are the necessity of educating consumers (in human biotechnology on vaccines, diagnostic kits, biomaterials or other new products) and strong patent protection (Arthur, 1994; Niosi, 2000). Thus the question may be asked whether increasing returns will transform this entrepreneurial, Schumpeter Mark I activity into a more mature, corporate, Schumpeter Mark II one. Are we witnessing the same kind of process that created dominant firms such as Microsoft in software, Boeing and an Airbus in the production of aircraft or Intel in micro-processors? Some data suggest that we are. Only a handful of DBFs in each industrial country are now profitable, most of them being in human health. Both Mangematin and Niosi suggest that a few biotechnology firms in each France and Canada are reaping the benefits of high growth, and positive benefits. These firms may prove the seeds of future concentration of biotechnology into a reduced number of larger, international corporations. Human therapeutics seems to be the area where the future winners will appear. A few critical success factors, such as a good patent strategy, venture capital support, well-conceived and well-timed R&D and marketing alliances, and low resistance by consumers, may propel a handful of biopharmaceutical firms towards industry dominance in the foreseeable future.

Biotechnology is a young and promising field. We have only started to witness its possibilities and impacts (both positive and negative). This book has brought together a few of the major issues and some of the authors that have contributed to the early analysis of the new domain. This conclusion only intends to emphasize that most of the work is in front of us. The volume will have succeeded if it has brought a few responses but many new interesting questions and if it has at the same time enlightened future paths for economic and social research.

NOTES

Chapter 2

[1] This paper is based on an early draft of chapter 2 (by Bo Carlsson, Staffan Jacobsson, Magnus Holmén, Annika Rickne, and Rikard Stankiewicz) in Bo Carlsson (ed.), New Technological Systems in the Bio Industries: An International Study (Boston and Dordrecht, Kluwer Academic Publishers, forthcoming). It draws on research within the project Sweden's Technological Systems and Future Development Potential. The project group consists of researchers at The Royal Institute of Technology in Stockholm, Chalmers University of Technology in Gothenburg, the Research Policy Institute at the University of Lund, and the Research Institute of Industrial Economics in Stockholm. The project is directed by Bo Carlsson at Case Western Reserve University in Cleveland, Ohio, and the project is funded by the Swedish National Board for Industrial and Technical Development (Nutek).

[2] A system may be defined as "a set or arrangement of things so related or connected as to form a unity or organic whole" (Webster's Collegiate Dictionary). Systems engineers define a system as a set of interrelated components working toward a common objective. Systems are made up of components, relationships, and attributes. Components are the operating parts of a system; relationships are the links between the components; and attributes are the properties of the components and the relationships between them. The problem with this definition for our purpose is that the systems we are interested in are much too fuzzy. The 'common objective' is often not clear, and the components, relationships, and attributes hard to define. Nevertheless, a systems view is adopted for reasons that will be made clear below.

[3] For example, the exaggerated emphasis on the role of "revolutions" and other discontinuities. This conflicts with the eclectic and highly cumulative nature of "technology as knowledge". The concept of discontinuity, in fact, applies best at the level of artifact.

[4] However, these measures are quite aggregated and provided little assistance when judging whether or not, for instance, microwave antenna technology is within the same knowledge field as optical and radio frequency technologies or microwave-related components (Holmén and Jacobsson 1998).

[5] At least this is true if we ignore the inherently somewhat arbitrary nature of industrial classifications. Some classifications are based on type of product, others on technology, and yet others on type of inputs. Sometimes all three criteria appear in the same system.

[6] Such as variance in the propensity to patent between firms and the difficulties to map technological activities in software by using patents.

[7] As Porter (1998 p. 202) puts it: "Drawing cluster boundaries is often a matter of degree, and involves a creative process informed by understanding the most important linkages and complementarities across industries and institutions to competition. The strength of these 'spill-overs' and their importance to productivity and innovation determine the ultimate boundaries."

[8] It was a superior performance in Sweden which we aimed to explain using a system approach (for a summary see Carlsson and Jacobsson 1997).

Chapter 3

[1] Paper prepared for the workshop The Economic and Social Dynamics of Biotechnology, organized by PRIME and Statistics Canada, Ottawa, February 24th and 25th, 2000.

[2] Adopting, thereby, the notion of state space popularized by the philosophers of the age of enlightenment. David Hume and John Locke discussed the world in terms of (1) memory, (2) logics and (3) imagination. Leibnitz, however, accepted no more imagination than all possible logical permutations of facts ("memory"; see Eliasson 1996, pp. 16f), the reducibility presumption. This definition is sufficient for our purpose.

[3] Please note that this places our model in complete contrast to the standard neoclassical model or its derivative, new growth theory in which actors are assumed (for the sake of nice aggregation) to be only marginally ignorant, missing information being easily available at a predetermined cost.

[4] As will be argued below, technology is a much too varied and unstable "unit" by which to define an industry, leaving demand or the customer as the only provider of reasonably stable theoretical characteristics for an industry definition.

[5] And entrepreneurs have to believe in being a winner to be called an entrepreneur (Eliasson 1990a, p. 283, 1996, p. 56).

[6] The venture capitalists also contribute managerial, financing and marketing competence through their network, but this comes after the identification.

[7] As modeled in the Swedish micro-to-macro model MOSES (Eliasson 1977, 1978, 1991).

[8] This section and the next draw heavily on Eliasson, Åsa (2000).

[9] Pharmacia merged with UpJohn in the 1990s. Pharmacia & UpJohn is currently merging with Monsanto into Pharmacia.

[10] Swedish-Finnish Stora Enso, the largest forest company in the world, however, has recently (Dagens Nyheter, March 12, 2000, p. A10) decided to terminate experimenting with GMO-trees. The reasons, they say, are only “ethical”. They think traditional cross breeding techniques will do as well. On the doubtful logics and morale of such ethical reasoning, see Eliasson, Å., 2000).

[11] Note that in the medium and low risk strategies the large firms ultimately carry the venture financing competence and risks. Å. Eliasson (2000) observes that this role of the large firms appears to be more important in agriculture and in food industries than in the health care industry.

[12] Thus, for instance, a deficient local venture capital industry is to be expected (Eliasson 1997b) if a related type of industry does not exist. The reason is that sophisticated venture capital competence is always based on related industrial experience.

[13] Even though the transition has not been sudden and direct, but by way of electromechanical and electronic analogue technologies.

[14] This logic is still very present in the macro R&D production functions (producing innovations) of so called new growth theory. See for instance Romer (1990).

[15] The other scare was MITI’s focus on “super computer” technology and Artificial intelligence to beat IBM in the race for the “fifth generation computing” (Eliasson 1996, pp. 180f). IBM at the same time was attempting in several failed approaches to integrate computing and communications (C&C) technology. This integration became the fifth generation computing.

[16] This model has been elaborated by many, for instance by Malinvaud 1967. It has been superimposed on innovation modelling by way of Arrow 1962 and has resulted in the linear version II of Schumpeter (1942), formulated, for instance, by Freeman (1974).

Chapter 4

[1] Personal communication from Maryanna Henkart, Division Director, Molecular and Cellular Biosciences, National Science Foundation, October 1999. These figures suggest that US expenditure for the five years 1994-1998 inclusive was at least $20 billion.

[2] SmithKline Beecham is to donate £1 billion over the next 20 years to conquer elephantiasis, a tropical disease. It will supply drugs free of charge to populations in

developing countries at risk of contracting the disease. The three drugs to be donated no longer have patent protection and cost less than 10p per dose (Mihill, 1998). The Bill & Melinda Gates Foundation is also donating $750M over 5 years to immunize children in developing countries against major killer diseases. This work will be carried out in conjunction with the World Health Organization, UNICEF and The World Bank (http://www.gatesfoundations.org). Such philanthropy is unlikely to extend to new drugs with high earning potential.

Chapter 6

[1] MERIT (Maastricht Economic Research Institute on Innovation and Technology), PO Box 616, 6200 MD Maastricht, The Netherlands. E-mail: a.arundel@merit.unimaas.nl.

[2] Ditner J-M, Lemarie S. What can we learn about the development of GMOs from the American and European field tests databases? Paper for the ICABR conference "The shape of the coming agricultural biotechnology transformation: Strategic investment and policy approaches from an economics perspective", University of Rome, June 17-19, 1999.

[3] Analyses of the field trial data for Europe is by Matthias Hocke of MERIT. Complete data for 1999 is not yet available, but the field trial counts up to July 1999 indicate that the number of trials should remain below the 1997 peak.

[4] Based on an analysis of the OECD STAN data for the mid 1990s. The share of pharmaceutical employment in total manufacturing employment, although rising in Europe, is unlikely to have exceeded 2% by 2000.

[5] Another factor is temporary and is due to the vintage nature of installed physical plant. Firms do not replace their production systems at the first sign of a new technology, but wait until it is financially advantageous to replace existing capital. This can further delay the introduction of new process technology.

[6] Interviews by MERIT with paper firms in the Netherlands suggests that biotechnology will suffer a comparative economic disadvantage for some time. The main benefit offered by biotechnology is that it could significantly decrease the environmental impacts of pulp and paper production. Stricter environmental regulation could be required to shift the existing technical trajectory based on chemicals and energy-intensive mechanical processes to biotechnology-based processes.

[7] This excludes 48 articles in which biotechnology is a minor issue.

[8] The Survey of Biotechnology Use in Canadian Industries-1996 includes 22 different biotechnologies, while the Biotechnology Firm Survey - 1977 includes 17 definitions. The experience of Statistics Canada with these surveys has led to a new

OECD committee on the collection of biotechnology statistics. See the article in this volume by Antoine Rose.

9 Many DBFs are not included because of the requirement for annual sales of 5 million or more, whereas many DBFs have no sales at all.

10 Employee-weighting adjusts for differences by firm in the number of employees. For example, a large firm with 1000 employees will contribute 10 times more to the average than a smaller firm with 100 employees. For results by firm, see Arundel and Rose (1998).

11 Analysis by Matthias Hocke of MERIT.

Chapter 8

1 Intersociety Working Group, American Association for the Advancement of Science (1997). Table I-12 shows colleges and universities conducted 8.9% of total US R&D (defense and non-defense) by dollar value in 1970, and 12.2% in 1996. Federal R&D went from 15.6% to 8.8% during the same period. Industry R&D rose from 69.1% in 1970 to 72.8% in 1996.

2 Mansfield surveyed R&D executives from 66 firms in seven industries. He found that recent academic research was cited as key to the development of 27% of new pharmaceutical products introduced from 1975-1985 (Mansfield 1994). His survey results show that researchers cited by firms for the importance of their discoveries typically receive both private and public funds for their research, with public funds outweighing those from industry by a factor of 3.

3 One underlying issue is the data constraint; that is, data collection is expensive so analysts frequently rely on the data that government agencies have collected for other purposes. The level of aggregation, definition of technologies or industrial sectors, and range of variables for which data is available often preclude direct analysis of the phenomena of interest. See, e.g., Griliches (1994).

4 See, e.g., Burrill & Company LLC. "Biotech 99 Life Sciences into the Millennium: The Biotechnology Industry Annual Report." San Francisco. 1999.

5 For an excellent discussion of issues relating to the use of patents as economic indicators, see Griliches (1990, 1992).

6 See e.g., Zucker, Darby and Brewer (1994), Zucker, Darby and Armstrong (1996), Zucker, Darby and Brewer (1998)

7 An alternative approach, utilized by Arundel and Rose (1999), among others, recognizes that biotechnology is a set of techniques, rather than an industry sector akin to e.g., pharmaceuticals or agrochemicals. An interesting extension of this study

would be to create a dataset of companies utilizing biotechnology anywhere in their operations. This approach would produce a larger, more diverse set of target companies, but create complexity in biotech-specific data collection.

[8] In their 1994 study, Zucker et al focus on the use of recombinant DNA technology for the development of human therapeutics, a subset of the activities we include in this study.

[9] Eliasson (1996) notes that the definition of biotechnology is empirically based, and "dependent on particular… techniques that constantly change." (pp. 13)

[10] This category consists of those who are designated "founder" or "co-founder" in company documents or official statements. It has been argued that this designation properly should be extended to initial scientific advisors, for example, or others whose expertise was crucial to successful establishment of the firm. A broader interpretation would likely increase the number of firms for which UC scientists could be considered founders; this issue will be taken up more exhaustively in later work.

[11] Biotechnology companies in the Critical Linkages database 1995-96

[12] See Pisano (1988) for a study of the role of strategic alliances between established firms and biotech start-ups, based on a similar line of argument.

[13] Including contracts for clinical research brings the total to 40 companies and 155 projects.

Chapter 10

[1] Anyone using the concept of regimes must do so knowing that there are varying usages of the term, including differences between its use in domestic versus international politics and policy. For example, Harris and Milkis view a regime as "a constellation of (1) new ideas justifying governmental control over business activity, (2) new institutions that structure regulatory politics, and (3) a new set of policies impinging on business" (Harris and Milkis, 1989, p. 25). They then use this concept of regime to examine the changes in health and safety regulation in the U.S. by the Reagan Administration compared to the earlier progressive era of US regulatory politics. It is important to note how regime has been used in the study of international relations. Here it is often used to convey softer arrangements and agreements among countries where there often is no real organization at its centre (Krasner, 1983; Zacker, 1996). Thus there are regimes for whaling or for Arctic pollution or any number of problems or issues up for discussion and action. Competition policy and regulation at the international level has been characterized as being more regime-like in this sense, because relationships tend to fall largely in the realm of notifications and information sharing among national or regional entities rather than in the realm of institutionalized dispute settlement through a core international organization (Doern and Wilks, 1996).

[2] For example, a 1998 Canadian summary account of international aspects of biotechnology included over 20 different kinds of international rules and influences emanating from various trade, environmental and food and health arenas of law and policy (Canada, 1998).

Chapter 11

[1] This may lead them closer to sociologists, especially ones involved in network theory and network relationships for economic processes (Powell et al 1996).

[2] This in turn, of course, relates to work by some business economists and economists.

[3] The reason that economists examine exports in addition to production is that international trade theory assumes that countries produce and trade those goods and services where they are relatively efficient, e.g. have a relative advantage. Hence, the relative specialization of a country's exports indicates which goods and services sectors that a national economy is relatively more efficient at producing compared to its trading partners

[4] Note, however, that while economic development and small companies may be an explicit and assumed outcome of investment in basic science, the question of 'how and why' this will occur has not always been clarified in detail.

[5] In analyzing these differences, however, questions arise about the accuracy of any comparative figures not gathered by national statistical agencies, including Ernst & Young (E&Y) biotechnology figures. E&Y do give explicit definitions of biotechnology and 'core biotechnology' companies, although there are always, latitudes of interpretive flexibility. Questions arise about every E&Y biotechnology report, whether from the USA or Europe or Australia. For example, their figures may easily be half or double the figures found by national researchers, and in some cases, the companies they identify as biotechnology firms may themselves say that they do not develop biotechnology.). The point here is not to question E&Y per se. Their reports are valuable for some purposes. The point is instead to highlight the current difficulties of cross-country comparisons of biotechnology and to call for internationally comparative definitions and statistics, potentially through the work of the OECD.

[6] In these sectors, the average size of Australian-owned firms is only one-tenth of foreign-owned firms. ISR (1999c: 32) argues that 'in medium-low and low R&D intensity industries, in which about 70 per cent of Australian manufacturing is concentrated, the dominance of foreign firms is much reduced.' The argument is, however, a relative rather than absolute statement. These Australian owned firms in medium to low R&D intensive sectors do more export and R&D than their counterparts in high tech sectors, but the figures say nothing about the

competitiveness of the Australian-owned firms in domestic or international markets.

7 Although Australia is leading in some environmental knowledge fields like photovoltaics, these areas do not appear to involve modern biotechnology.
8 This is not the same thing as saying, however, that every major university with research in the area has a cluster of biotechnology firms!

Chapter 12

1 As the mortality rate is high, our figure under-estimates the number of firms set up during the last 10 years. But it gives an idea of the number of firms set up and which survive.

2 This figure does not take into account divisions in certain firms specialised in biotechnology, with over 500 employees.

3 Ministère de l'Education Nationale, de la Recherche et de la Technologie. The French ministry in charge of research.

4 Sessi, 1999: "La situation de l'industrie", Annual Business Survey, Sessi, Paris.

5 Consultancy research organisation

6 thousand euros

7 million euros

8 Since there is no significant difference in the set-up dates of the different groups, it seems likely that venture capital invests little in firms in this group, for its absence cannot be imputed to a withdrawal of venture capital companies after the sale of their shares.

9 In some cases it is difficult to identify patents since some were registered in the name of the inventor before being incorporated into the company's assets.

10 A steering committee composed of Pascale Auroy (ARD), Christine Bagnaro (ANVAR), Patrice Blanchet (DTA/2, MENRT), Marie José Dudézert (DTA/2, MENRT), Anne Sophie Godon (Arthur Andersen) and Vincent Mangematin (INRA/SERD) met in 1998-99 under the chairmanship of Jean Alexis Grimaud, to plan and carry out the survey.

Chapter 13

1 I acknowledge with thanks the helpful and constructive comments on an earlier draft I have received from Anthony Arundel, Susan McDaniel, Karol J, Krotki, Volker Meja, Michael Smith, Hermann Strasser and Jay Weinstein. I am grateful to

Paul Malone for his valuable editorial assistance. Email: Nico.Stehr@gkss.de

[2] "Back to basics", The Guardian, February 1, 2000.

[3] See "Biotech fuelling latest revolution, economist says", Globe and Mail Wednesday, November 24, 1999.

[4] A differentiation between biotechnology products is not only advisable because consumers, for the time being at least, as I would emphasize, appear to be ready to asses biotechnology products within the field of medicine more favorably. Public discourse about such potential pharmaceuticals and GM food products therefore products differ significantly (see Conrad, 1999). One needs to distinguish here between "promised" biotechnology products mainly in the field of medical genetics where discourse is virtually controlled its proponents and products that already are competing in the marketplace. There is no guarantee that discourse about medical genetics will continue along the same path (cf. "Financial ties in biomedicine get close look", New York Times, February 20, 2000).

[5] The terminology in biotechnology is quite fluid. Terms are used interchangeably and there is a vigorous contest to gain domination in the meanings assigned to different concepts such as genetics, genetic manipulation, agricultural genetics, genetic engineering, genetic modification, new biotechnology. It matters if one initially assigns and discusses biotechnology products with respect field such as "disease, health and medicine" or "food" although the two domains are clearly related.

[6] Fred Hassan, the chief executive of Upjohn, the drugs groups, had rather bullish comments about the prospect of a merger between his company and Monsanto, the US biotechnology company although he admitted that an "education" campaign was needed to overcome the "PR problem" of the biotechnology industry. In January of 2000, however, Hassan of Upjohn offered to axe a $ 600 million research program for genetically modified foods "in an attempt to pacify shareholder unhappy about its takeover of Monsanto" ("P&U offers to scrap $ 600m GM plan research", The Guardian, February 1, 2000). The new, merged company would also, Hassan indicated, drop the Monsanto name. The company will be called Pharmacia if investors approve the merger.

[7] Compare the article by Floyd Norris "Public misinformed on genetically modified foods" (New York Times, December, 17, 1999) in which he argues that companies such as Monsanto made a serious strategic error in not targeting or persuading the public instead of farmers and the Food and Drug Administration about the "evidently small risks" that offset the potential gains of genetically modified seeds.

[8] The term "market behavior" is designed to signal that I also refer to consumer decisions and reasons for decisions in the market place when I employ the term "consumption". This usage of the term consumption differs from that advanced by

Douglas and Isherwood (1979:57; emphasis added) who refer to "consumption as a use of material possessions that is beyond commerce and free within the law."

[9] Max Weber ([1922] 1978:636) has described the sober, distanced and indirect personal relations that characterize of the ideal-typical modern market place resulting from such an interest or utility-based orientation to the market quite well: "The market community as such is the most impersonal form of practical life into which humans can enter with one another.... The reason for the impersonality of the market is its matter-of-factness, its orientation to the commodity and only to that. When the market is allowed to follow its own autonomous tendencies, its participants do not look toward the persons of each other but only toward the commodity, there are no obligations of brotherliness or reverence, and none of these spontaneous human relations that are sustained by personal unions."

[10] The perhaps classical difference or counter concepts challenging the system specific autonomy of the economic market are the terms "plan" or "state". In the context of this paper I do not intend to confront market-plan-state relations.

[11] The assertion that we are moving toward markets where decisions of market participants are increasingly based on a moralization of economic action may be criticized on at least two counts: (1) The economic system and economic action were really never all that disattached from other systems and forms of discourse. (2) More recently, the sustainable development debate, the demand for equal opportunities, health and safety regulations or, the co-determination legislation has already gone a long way to couple economic action, assuming it operated more or less according to its own logic, to moral discourse and political purposes.

[12] The assertion that the modern economic system is moving, based on endogenous developments, from "production-orientation" to "consumption-orientation" indeed has been advanced repeatedly in the last century, usually during periods of sustained economic growth. But these debates were quite different from the economic transformations analyzed in this paper. Claims about the end of scarcity, the declining need of capitalism for asceticism and the rise of virtues surrounding the act of consumption were heard in the United States in the 1920s and after World War II. However, the discussion in each case centered on the issue of expanding the volume of consumption and its cultural consequences. In the words of John K. Galbraith ([1967] 1971:37), the "individual serves the industrial system not by supplying it with savings and the resulting capital; he serves it by consuming its products." A critiques of these perspectives may be found in Martin, 1999. Discussions that centered on the need to consume soon were followed by treatise after treatise informed by moral indignation about consumerism; these works, castigating consumers as vulgar, greedy, stupid and insentitive, reinvoking Thorstein Veblen's scorn of conspicuous consumption (see Douglas and Isherwood, 1979:vii-viii).

[13] I have borrowed the title of this section from Joseph A. Schumpeter's Capitalism, Socialism and Democracy ([1942] 1962). Schumpeter, although highly attuned to the

evolutionary character of the capitalist system, is, as is well-known, rather pessimistic about the long-term dynamics of the capitalist economy. It is bound to fail because it becomes too successful, it breaks "to pieces under the pressure of its own success" (Schumpeter, [1942] 1962: 134). Both Marx and Schumpeter have, in the end, underestimated the ability of the capitalist system to evolve beyond its allegedly inherent limits. Schumpeter's ([1942] 1962:125) title suggests something else, however; a thoroughly Marxian imagery, namely that "all the features and achievements of modern civilization are, directly or indirectly, the products of the capitalist process". Modern society is not --to mention but one alternative -- politically constituted. The concept of society becomes economic. If one did not care to accept Schumpeter's sweeping assertion, more appropriate title of this section would be "civilizing capitalism". A similar general premise about the seizure of control of society and its institutions by capitalism surfaces in the context of critical theory and elsewhere in social theory. Most recently, this view finds expression in the idea a colonization of the life-world by the systems of capitalism, bureaucracy and law (Habermas, 1981, 1987).

[14] In his suggestive system-theoretic perspective of economic activities, Niklas Luhmann (1988:94) defines the "market" as the system internal environment of the economic subsystems. The economic system creates its own internal environment in order to achieve reductions of complexity for the purpose of observing external environments such as the state, scientific and technological developments or ecological transformations.

[15] Traditional social thought champions centrality and unity: Aside from a strictly system theoretic perspective, the idea as the title "The civilization of capitalism" clearly signals, that the logic or rational attitude of economic affairs would subjugate – rationalize – almost every forms of life external to the economy has for the longest period hardly been a contested assumption in social theory. While a system theoretic perspective (Luhmann) resist assigning overall functional priority to any social system and the exclusive logic that governs any subsystem, the much more common premise is that the economy is the dominant social system in modern societies. Traffic that is seen to occur in the opposite direction represents a controversial perspective and that the assumed rationalization of life styles, social institutions, ideas etc. has been less than complete also is contentious assertion.

[16] The emergence of market as a separate social institution is seen by prominent economic historians as benign or even benevolent development. Albert Hirschman (1977) in his book The Passions and the Interests has argued forcefully that the pursuit of "interest" in the context of the evolving market was a vast improvement on the pursuit of "passion" since interest was governed by reason rather unruly emotions. Not only radical opponents of market economies in the past but also more liberal observers at the present time take, with respect to the consequences for the moral fabric of society, exactly the opposite position: "the market contributes more to the erosion of our moral sense than any other modern social force" (Schwartz, 1999:37).

[17] The translation of knowledge claims or the "pragmatization" of scientific and technical knowledge is the job performed by experts, counselors and advisors who apply knowledge to knowledge. Counselors, advisers, and experts who are needed to mediate between the complex distribution of changing knowledge and those who search for enabling knowledge. Ideas travel and are transferred as people's "baggage" (that is, knowledge is "encultured", "embrained" or "encoded"), whereas skills (in the sense of know-how or rules of thumb) are more firmly inscribed, embedded or embodied in people, objects and resources. Rather than seeing knowledge as something that is arrested and that is there, knowing should be seen as an activity, as something that people do. The active intervention is of particular relevance in the case of specialized scientific knowledge that is, as I have stressed, contestable and often de-pragmatized. A chain of interpretations, or the essential "openness" of knowledge claims, must come to an "end" in order to become relevant in practical contexts and therefore effective as a capacity of action. Experts in modern society largely perform this function of ending reflection, of reducing the openness and contestability of knowledge for the purpose of action. But what is new is not that knowledge-based work and workplaces that require such cognitive skills are emerging. Experts have always been around. The significant economic transition in general and transformation of the labor market in particular are to be found in the number and proportion of workplaces that require knowledge-based work and the in decline of workplaces that make or move things.

[18] Touraine ([1992] 1995:207) captures the increased value of individualism in contemporary society and, in its wake, the rise of consumerism as follows: "The modern world … increasingly abounds with references to a Subject. That Subject is freedom, and the criterion of the good is the individual's ability to control his or her actions and situation, to see and experience modes of behavior as components in a personal life history, to see himself or herself as an actor. The Subject is an individual's will to act and to be recognised as an actor."

[19] My observation about the role of consumption indicates that there evidently is an elective affinity to Bourdieu's (1984) examination of styles of consumption or consumption as competitive display, in his study Distinction: A Social Critique of the Judgment of Taste. Consumption is more than the mere deployment of economic resources or the satisfaction of material welfare. Styles of consumption exhibit differential "cultural capital". The pageant of products is part of an unequal system of reputation and distinction (see also Friedman, 1994; Bauman, 1998).

[20] According to Anthony Arundel (personal communication) the "'modest pockets of non-biotechnology' constitute ninety percent of the research effort by European seed firms today and an estimated eighty five percent in 2002. Buttel has apparently fallen hook, line and sinker for the continual propaganda from the biotechnology government-industry complex about the economic and technical importance of biotechnology;" see also Arundel, 2000.

[21] The source of the standards chosen to police knowledge, the regulatory procedures put in place, and the intellectual systems legitimizing the cultural dismissal of certain uses of knowledge typically also do not originate in science and technology itself. For example, in the face of demands to preserve and defend the nature of human nature in response to developments in scientific and technical capacities to alter the status quo of human reproduction, scientific 'notions of nature do not provide us with unambiguous standards of naturalness to which we can appeal for normative orientation' (van den Daele, 1992: 549). Since scientific notions of naturalness allow for the construction of a range of possible natures, regulation efforts advancing the cause of abstaining from practical steps intervening into human nature have to appeal to moral claims and political action that may or may not succeed in arresting human nature. The anchoring of standards and justifications outside of science does not mean that individuals who are scientists may not be found among those who vigorously support attempts to regulate knowledge.

[22] A German study (Gath and Alvensleben, 1997) concurs that the "possibilities of influencing the acceptance of GM foods by information are limited…the widespread opinion that the acceptance of GM food is primarily an information and education problem … has to be questioned."

[23] I have to leave open another, perhaps even more fascinating issue that addresses the relationships between active citizens and governance in modern society. Rose (1999:166) refers to a symbiosis between more sovereign citizens and forms of modern government when he observers that "advanced liberal forms of government … rest, in new ways, upon the activation of the powers of the citizen." And the new, active role of the citizen includes the citizen as a consumer who becomes an active agent, not just active in the dynamics of markets transacting products but "in the regulation of professional expertise" (Rose, 1999:166) or symbolic properties.

BIBLIOGRAPHY

Abbott, F. M. (1989). "Protecting First World Assets in the Third World: Intellectual Property Negotiations in the GATT Multilateral Framework", *Vanderbilt Journal of Transnational Law*, Vol. 22, No. 4, pp. 689-746.

Abramovitz, M. (1988). "Thinking about Growth", in Abramovitz, M. (ed.), *Thinking about Growth*, Cambridge: Cambridge University Press.

Acharya, R. (1992). "Patenting of Biotechnology: GATT and the Erosion of the World's Biodiversity", *Journal of World Trade*, Vol. 25, No. 6, pp. 71-87.

Acharya, R., Arundel, A. and Orsenigo, L. (1998). "The evolution of European biotechnology and its future competitiveness", in Senker, J. (ed.) *Biotechnology and Competitive Advantage: Europe's Firms and the US Challenge*, Northampton: Edward Elgar.

Acharya, R., Ziesemer, T. (1996). "A closed economy model of horizontal and vertical product differentiation: The case of innovation in biotechnology", *Economics of Innovation and New Technology*, Vol. 4, pp. 245-264.

ACOST (1990). "Developments in Biotechnology", London: HMSO.

Adams, J. (1995). *Risk*, London: UCL Press.

Alexander, G. (2000). "Technology Bubble Hits Bursting Point", *Sunday Times*, March 19, Section 3, p. 10.

Anderson, L. (1999). *Genetic Engineering, Food and Our Environment*, Dartington: Green Books.

Appleyard, B. (1999). *Brave New Worlds*, London: Harper Collins.

Arora, A. and Gambardella, A. (1990). "Complementarity and external linkages: the strategies of the large firms in biotechnology", *The Journal of Industrial Economics*, Vol. XXXVIII, No. 4, pp. 361-380.

Arrègle, J. L. (1996). "Analyse resource Based et identification des actifs stratégiques", *Revue Française de Gestion*, (Mars-avril-mai), pp. 25-35.

Arrow, K. J. (1962). "Economic Welfare and the Allocation of Resources for Invention", in Nelson, R. (ed.), The *Rate and Direction of Inventive Activity: Economic and Social Factors*, Princeton: NBER, Princeton University Press.

Arthur, W. B. (1994). *Increasing Returns and Path Dependence in the Economy*, Ann Arbor: University of Michigan Press.

Arundel, A. (1999). "Diffusion of biotechnologies in Canada", Research Paper No. 6, Science and Technology Redesign Project, Statistics Canada publication 88F0017MPB.

Arundel, A. (2000). "Measuring the Use and Planned use of Biotechnologies by firms", paper presented at the PRIME/Statistics Canada Advanced Research Workshop on "The Economic and Social Dynamics of Biotechnology", Ottawa, Ontario, Canada, February 24-25.

Arundel, A. and Rose, A. (1999a). "Employment crisis in the making?", *Biotech*, June/July, pp. 17-18.

Arundel, A. and Rose, A. (1999b). "The diffusion of environmental biotechnology in Canada: adoption strategies and cost offsets", *Technovation*, Vol. 19, 551-560.

Ashford, T. (1996). "Regulating Agricultural Biotechnology: Reflexive Modernization and the European Union", *Policy and Politics*, Vol. 24, No. 2 (April), pp. 125-136.

Audretsch, D. and Stephan, P. (1996). "Company scientist locational links: the case of biotechnology", *American Economic Review*, Vol. 86, No. 3, pp. 641-652.

Baker, C. (2000). "Factors Influencing Public Confidence From Environics International Survey on Food Safety and Biotechnology", paper presented to the Workshop on Public Confidence in Biotechnology, Ottawa, March 4.

Barbanti, P., Gambardella, A. and Orsenigo, L. (1999). "The Evolution of Collaborative Relationships Among Firms in Biotechnology", *International Journal of Biotechnology*, Vol. 1, No. 1, pp. 10-29.

Barley, S. R., Freeman, J. H. and Hybels, R. C. (1992). "Strategic Alliances in Commercial Biotechnology", in Nohria, N. and Eccles, R. G. (eds) *Networks and Organizations*, Boston: Harvard Business School Press.

Barnes, B. (1999). "Biotechnology as Expertise", in O'Mahony, P. (ed.) *Nature, Risk and Responsibility Discourses of Biotechnology*, New York: Routledge.

Barney, J. B. (1991). "Firm resources and sustained competitive advantage", *Journal of Management*, Vol. 17, pp. 99-120.

Barney, J. B., Spender, J. C., and Reve, T. (eds) (1994). "Does management matter? On competencies and competitive advantage", Crafoord Lectures, Vol. 6. Lund.

Batholomew, S. (1997). "National Systems of Biotechnology Innovation: Complex Interdependence in the Global System", *Journal of International Business Studies*, Vol. 28, No. 2, pp. 241-66.

Bauman, Z. (1998). *Work, Consumerism and the New Poor*, Buckingham: Open University Press.

Beck, U. (1992). *Risk Society: Towards a New Modernity*, London: Sage.

Benbrook, C. (1999). "Evidence of the magnitude and consequences of the Roundup Ready soybean yield drag from University-based varietal trials in 1998", Agbiotech InfoNet Technical Paper No. 1, July.

Bhat, M. G. (1996). "Trade Related Intellectual Property Rights To Biological Research - Socioeconomic Implications for Developing Countries", *Ecological Economics*, Vol. 19, No. 3, pp. 205-17.

Bijker, W. E. (1995). *Of Bicycles, Bakelites and Bulbs. Towards a Theory of Sociotechnical Change (Inside Technology)*, Cambridge, MA: MIT Press.

Bijman, J. (1999). "Life science companies: Can they combine seeds, agrochemicals and pharmaceuticals", *Biotechnology and Development Monitor*, Vol. 40, pp.14-19.

BioNews available from <BioNews@lists.progress.org.uk>

BIOTECanada (1998). *Canadian Biotechnology 98: Success from Excellence*, BIOTECanada, Ottawa: BIOTECanada.

Biotechnology Research Subcommittee, Committee on Fundamental Science, National Science and Technology Council (1995). "Biotechnology for the 21st Century: New Horizons", Washington: US Government Printing Office.

Bourdieu, P. (1984). *Distinction: A Social Critique of the Judgment of Taste*, London: Routledge & Kegan Paul.

Breschi, S. and Malerba, F. (1997). "Sectoral Innovation Systems: Technological Regimes, Schumpeterian Dynamics and Spatial Boundaries", in Edquist, C. (ed.) *Systems of Innovation: Technologies, Institutions and Organizations*, London: Pinter Publishers.

Browne, D. (with Chaitoo, R. and Hart, M.) (2000). "Can Eco-Labelling Undermine International Agreement on Science-Based Standards?", in Doern G. B. and Reed, T. (eds) *Risky Business: Canada's Changing Science-Based Policy and Regulatory Regime*, Toronto: University of Toronto Press.

Bull, A. T., Holt, G. and Lilly, M. D. (1982). *Biotechnology, International Trends and Perspectives*, Paris: OECD.

Burke, J. F. and Thomas, S. M. (1997). "Agriculture is biotechnology's future in Europe", *Nature Biotechnology*, Vol. 15, p. 695.

Buttel, F. H. (1999). "Agricultural biotechnology: it recent evolution and implications for agrofood political economy", *Sociological Research Online*, <www.socresonline.org.uk/socresonline/-1995/1/buttel.html>

Buttel, F. H. (2000). "The recombinant BGH controversy in the United States: toward a new consumption politics of food?", *Agriculture and Human Values*, Vol. 17, No. 1.

Buttel, F. H., Cowan, J. T., Kenney, M. and Kloppenburg Jr., J. (1984). "Biotechnology in Agriculture: The Political Economy of Agribusiness Reorganization and Industry-University Relationships", *Research in Rural Sociology and Development*, Vol. 1, pp. 315-348.

Callon, M. (1999). "Actor network theory – the market test", in Law, J. and Hassard, J. (eds) *Actor Network Theory and After*, Oxford: Blackwell.

Callon, M. and Law, J. (1989). "On the Construction of Sociotechnical Networks; Content and Context Revisited", *Knowledge and Society*, Vol. 9, pp. 57-83.

Callon, M. and Vignolle, J. (1976). "Organisation locale et enjeux sociétaux", *Sociologie du travail*, Vol. 3, No. 2, pp. 233-255.

Callon, M., Laredo, P., Rabeharisoa, V., Gonard, T. and Leray, T. (1992). "The Management and Evaluation of Technological Programs and the Dynamics of Technoeconomic Networks – The Case of the AFME", *Research Policy*, Vol. 21, No. 3, pp. 215-236.

Canada (1998). *Renewal of The Canadian Biotechnology Strategy: International Issues Report*, Ottawa: Industry Canada.

Canada, Commission of Inquiry on the Blood System of Canada (The Krever Inquiry) (1997). *Final Report*, Ottawa: Minister of Public Works and Government Services.

Canadian Biotech News Service: *Canadian Biotech News*, Nepean, Ont. (weekly)

Carlsson, B. (ed.) (1989). *Industrial Dynamics: Technological, Organizational, and Structural Changes in Industries and Firms*, Boston and Dordrecht: Kluwer Academic Publishers.

Carlsson, B. (ed.) (1995). *Technological Systems and Economic Performance: The Case of Factory Automation*, Boston, Dordrecht, London: Kluwer Academic Publishers.

Carlsson, B. (ed.) (1997). *Technological Systems and Industrial Dynamics*, Boston, Dordrecht and London: Kluwer Academic Publishers.

Carlsson, B. (ed.) (2000). *New Technological Systems in the Bio Industries: An International Study*, Boston, Dordrecht and London: Kluwer Academic Publishers, (forthcoming).

Carlsson, B. and Jacobsson, S. (1997). "In Search of a Useful Technology Policy - General Lessons and Key Issues for Policy Makers," in Carlsson, B. (ed.) *Technological Systems and Industrial Dynamics*, Boston, Dordrecht and London: Kluwer Academic Publishers.

Carlsson, B. and Stankiewicz, R. (1995): "On the Nature, Function and Composition of Technological Systems," in Carlsson, B. (ed.) *Technological Systems and Economic Performance: The Case of Factory Automation*, Boston, Dordrecht, London: Kluwer Academic Publishers.

Cassier, M. (1999). "Research contracts between university and industry: co-operation and hybridization between academic research industrial research", *International Journal of Biotechnology*, Vol. 1, No. 1, pp. 82-104.

Centre for Medicines Research International (1997). *1997 Annual Report*, Carshalton, Surrey: Centre for Medicines Research International.

Chemistry & Industry, No. 7, 5 April 1999, p. 251.

Coase, R. H. (1960). "The Problem of Social Cost", *Journal of Law & Economics*, Vol. 3, (Oct.), pp. 1-44.

Cockcroft, C. (1999). "How plants fight thirst", *Guardian Online*, 25 November 1999.

Cohen, S. N., Chang, A. C. Y., Boyer, H. W. and Helling, R. B. (1973). "Construction of biologically functional bacterial plasmids in vitro", *Proceedings of the National Academy of Sciences*, USA, Vol. 70, pp. 3240-3244.

Connelly, J. and Smith, G. (1999). *Politics and the Environment*, London: Routledge.

Conrad, P. (1999). "A mirage of genes", *Sociology of Health and Illness*, Vol. 21, pp. 228-241.

Contact Canada (1993, 1995, 1998, 1999). *Canadian Biotechnology Directory*, Ottawa.

Cornerhouse (1998). "Food? Health? Hope? Genetic Engineering and World Hunger", Briefing No. 10, The Corner House, Sturminster Newton, Dorset.

Dahmén, E. (1950). *Svensk industriell företagarverksamhet* (Swedish Industrial Entrepreneurial Activity), Stockholm: Industriens Utredningsinstitut.

Dahmén, E. (1989). "Development Blocks in Industrial Economics," in Carlsson, B. (ed.) *Industrial Dynamics: Technological, Organizational, and Structural Changes in Industries and Firms*, Boston and Dordrecht: Kluwer Academic Publishers.

Darby, M. R. and Zucker, L. (1996). "Star Scientists, Institutions and the Entry of Japanese Biotechnology Enterprises", Working Paper W5795, Cambridge MA: National Bureau of Economic Research.

Darby, M. R., Liu, Q. and Zucker, L. (1999). "Stakes and Stars: The Effect of Intellectual Human Capital on the Level and Variability of High Tech Firms'

Market Values", Working Paper W7201, Cambridge, MA: National Bureau of Economic Research.

DaSilva, E. J. and Ratledge, C. (eds) (1992). *Biotechnology: Economic and Social Aspects*, Cambridge: Cambridge University Press.

Davies, D. (ed.) (1986). *Industrial Biotechnology in Europe. Issues for Public Policy*, London: Pinter.

de la Mothe, J. (ed.) (2000). *Science, Technology and Governance*, London: Pinter.

de la Mothe, J. and Ducharme, L. M. (eds) (1990). *Science, Technology and Free Trade*, London: Pinter.

de la Mothe, J. and Paquet, G. (eds) (1996). *Evolutionary Economics and the New International Political Economy*, London: Pinter.

de la Mothe, J. and Paquet, G. (eds) (1998). *Local and Regional Systems of Innovation*, Boston: Kluwer.

de la Mothe, J. and Paquet, G. (eds) (1999). *Information, Innovation and Impacts*, Boston: Kluwer.

De Solla Price, D. (1984). "The science/technology relationship, the craft of experimental science, and policy for the improvement of high technology innovation", *Research Policy*, Vol. 13, No. 1, pp. 3-20.

Dodgson, M. (1993). *Technological Collaboration in Industry*, London: Routledge.

Doern, G. B. (1999). *Global Change and Intellectual Property Agencies.* London: Pinter.

Doern, G. B. (2000). "Biotechnology, Public Confidence and Governance", paper for the Workshop on Public Confidence in Biotechnology, Industry Canada, Ottawa, March 4.

Doern, G. B. and Reed, T. (eds) (2000). *Risky Business: Canada's Changing Science-Based Policy and Regulatory Regime*, Toronto: University of Toronto Press.

Doern, G. B. and Sharaput, M. (2000). *Canadian Intellectual Property: The Politics of Innovating Institutions and Interests*, Toronto: University of Toronto Press, (forthcoming).

Doern, G. B. and Sheehy, H. (1999). "The Federal Biotechnology Regulatory System: A Commentary on an Institutional Work in Progress", in Knoppers and Mathios, (eds) *Biotechnology and the Consumer*, Dordrecht: Kluwer.

Doern, G. B. and Wilks, S. (eds) (1996). *National Competition Policy Institutions in a Global Market*, Oxford: Clarendon.

Doremus, P. N. (1996). "The Externalization of Domestic Regulation - Intellectual Property Rights Reform in a Global Era", *Science Communication*, Vol. 17, No. 2, pp. 137-162.

Dosi, G. (1982). "Technological Paradigms and Technological Trajectories – A Suggested Interpretation of the Determinants and Directions of Technical Change", *Research Policy*, Vol. 11, No. 3, pp. 147-162.

Douglas, M. and Isherwood, B. (1979). *The World of Goods*, New York: Basic Books.

Drahos, P. (1996). "Global Law Reform and Rent-Seeking: The Case of Intellectual Property", *Australian Journal of Corporate Law*, Vol. 7, pp. 45-61.

Drahos, P. (1997). "Thinking Strategically about Intellectual Property Rights", *Telecommunications Policy*, Vol. 21, No. 3, pp. 201-11.

Economic Research Service (1999). "Impacts of adopting genetically engineered crops in the U.S.: Preliminary results. ERS", United States Department of Agriculture, www.econ.ag.gov/whatsnew/issues/gmo, July 20.

Ehrnberg, E. and Sjöberg, N. (1995). "Technological Discontinuities, Competition and Firm Performance", *Analysis & Strategic Management*, Vol. 7, No. 1, pp. 93-107.

Eliasson, Å. (2000). "A Competence Bloc Analysis of the Economic Potential of Biotechnology in Agriculture and Food Production", Working Paper IBMP-CNRS, Stockholm: Strasbourg and KTH.

Eliasson, G. (1977). "Competition and Market Processes in a Simulation Model of the Swedish Economy", *American Economic Review*, Vol. 67, No. 1, pp. 277-281.

Eliasson, G. (1978). "Micro-to-Macro Model of the Swedish Economy", Conference Reports, 1978:1. Stockholm: IUI.

Eliasson, G. (1987). "Technological Competition and Trade in the Experimentally Organized Economy", Research Report No. 32, Stockholm: IUI.

Eliasson, G. (1989). "The Dynamics of Supply and Economic Growth – how industrial knowledge accumulation drives a path-dependent economic process", in Carlsson, B. (ed.) *Industrial Dynamics: Technological, Organizational, and Structural Changes in Industries and Firms*, Boston and Dordrecht: Kluwer Academic Publishers.

Eliasson, G. (1990a). "The Firm as a Competent Team", *Journal of Economic Behavior and Organization*, Vol. 13, No. 3, pp. 275-298.

Eliasson, G. (1990b). "The Knowledge-Based Information Economy", in Eliasson, G., Fölster, S. et al., *The Knowledge Based Information Economy*, Stockholm: IUI.

Eliasson, G. (1991). "Modeling the Experimentally Organized Economy", *Journal of Economic Behavior and Organization*, Vol. 16, No. 1-2, pp. 153-182.

Eliasson, G. (1992). "Business Competence, Organizational Learning, and Economic Growth: Establishing the Smith-Schumpeter-Wicksell (SSW) Connection", in Scherer, F. M. and Perlman, M. (eds) *Entrepreneurship, Technological Innovation, and Economic Growth. Studies in the Schumpeterian Tradition*, Ann Arbor: The University of Michigan Press.

Eliasson, G. (1996). *Firm Objectives, Controls and Organization – the use of information and the transfer of knowledge within the firm*, Boston, Dordrecht, London: Kluwer Academic Publishers.

Eliasson, G. (1997a). "General Purpose Technologies, Industrial Competence Blocs and Economic Growth", in Carlsson, B. (ed.) *Technological Systems and Industrial Dynamics*, Boston, Dordrecht and London: Kluwer Academic Publishers.

Eliasson, G. (1997b). The Venture Capitalist as a Competent Outsider; mimeo, INDEK, KTH, IEO R: 1997-06. Stockholm.

Eliasson, G. (1998a). "Svensk datorindustri – en kompetensblocksanalys av dess framväxt och försvinnande" (Swedish Computer Industry – a competence bloc analysis of its emergence and disappearance), in Heum, P. (ed.) *Kompetense og Verdiskapning. SNFs Årsbok*, Bergen: Fagboksforlaget.

Eliasson, G. (1998b). "Information Efficiency, Production Organization and Systems Productivity - Quantifying the Systems Effects of EDI Investments", INDEK, KTH, TRITA-IEO R 1996:6; in Macdonald, Madden, Salama (eds) *Telecommunications and Social Economic Development*, Amsterdam: North Holland.

Eliasson, G. (1998c). "Industrial Policy, Competence Blocs and the Role of Science in the Economic Development", KTH, TRITA.IEO R 1998-08, *Journal of Evolutionary Economics*, No. 1, 2000 (forthcoming).

Eliasson, G. (1998d). The Nature of Economic Change and Management in the Knowledge-Based Information Economy. KTH TRIRA-IEO-R 1998:19.

Eliasson, G. and Eliasson, A. (1996). "The Biotechnological Competence Bloc", *Revue D'Economie Industrielle*, Vol. 78, No. 4, pp. 7-26.

Eliasson, G. and Eliasson, Å. (1997). "The Biotechnological and Pharmaceutical Competence Bloc", in Carlsson, B. (ed.) *Technological Systems and Industrial Dynamics*, Boston, Dordrecht and London: Kluwer Academic Publishers.

Eliasson, G. and Eliasson, Å. (2000). The Markets for Venture Capital, Strategic Acquisitions and Contract Work, mimeo, KTH, Stockholm.

Eliasson, G. and Taymaz, E. (1999). "Institutions, Entrepreneurship, Economic Flexibility and Growth – experiments on an evolutionary model", KTH, INDEK, TRITA-IEO-R 1999:13, to be published in Cantner, Hanush, Klepper, (eds) *Economic Evolution, Learning and Complexity – Econometric, Experimental and Simulation Approaches*.

Elliott, L. (1999). "Bank woos AIDS drug firms", *The Guardian*, 21 June, 21.

Environment Canada (2000). "Backgrounder on the Biosafety Protocol to the UN Convention on Biological Diversity", Ottawa: Environment Canada.

Enzing, C., Benedictus, J. and Engelen-Smeets, E.; Senker, J. and Martin, P.; Reiss, T. and Schmidt, H.; Assouline, G., Joly, P.B. and Nesta, L. (1999). *Inventory of Public Biotechnology R&D Programme in Europe*, Analytical Report, Luxembourg: Office for Official Publications of the European Communities, Vol. 1.

Ernst and Young (1999a). *Australian Biotechnology Report 1999, Commonwealth of Australia,,* Canberra..

Ernst and Young (1999b). *European Life Sciences 99,*Stuttgart: Ernst and Young International.

EuropaBio (1997). "Benchmarking the Competitiveness of Biotechnology in Europe", Brussels: EuropaBio.

European Commission (1998). *Guidelines on the Application of the Precautionary Principle*, Luxembourg: Office for Official Publications of the European Communities. (Brussels) (Draft paper).

European Commission DG XII (1997). *The European and Modern Biotechnology*, Eurobarometer 46.1, Luxembourg: Office for Official Publications of the European Communities.

Faulkner, W., Senker, J. and Velho, L. (1995). *Knowledge Frontiers. Industrial Innovation and Public Sector Research in Biotechnology*, Oxford: Clarendon Press.

Feldman, M. (1999). "The New Economics of Innovation, Spillover and Agglomeration: Review of Empirical Studies", *Economics of Innovation and New Technology*, Vol. 8, No. 1, pp. 5-25.

Fox, J. (1996). "WHO and UNICEF find vaccines too costly", *Biotechnology*, Vol. 14, No. 11, p. 1532.

Francis, J. (1993). *The Politics of Regulation. A Comparative Perspective*, Oxford: Basil Blackwell.

Freeman, C. (1974). *The Economics of Industrial Innovation*, Harmondsworth: Penguin.

Freeman, C. (1988). "Japan: A New National System of Innovation?", in Dosi, G. et al., *Technical Change and Economic Theory*, London: Pinter.

Freeman, C. and Perez, C. (1988). "Structural Crises of Adjustment: Business Cycles and Investment Behaviour", in Dosi, G. et al. (eds) *Technical Change and Economic Theory*, London and New York: Pinter.

Frenkel, S. J., Korcynski, M., Shire, K. A. and Tam, M. (1999). *On the Front Line. Organization of Work in the Information Economy*, Ithaca, New York: Cornell University Press.

Friedman, J. (ed.) (1994). *Consumption and Identity*, London: Harwood.

Friedrich, G. A. (1996). "Moving beyond the genome projects", *Nature Biotechnology*, Vol. 14, No. 19, pp. 1234-1237.

Galbraith, J. K. ([1967] 1971). *The New Industrial State*, Boston: Houghton Mifflin.

Gath, M. and Alvensleben, R. v. (1997). "The potential effects of labeling GM foods on consumer decisions", Preliminary Report, Institute for Agricultural Economics. Universität Kiel.

Genome Canada (1999). *Genome Canada Business Plan: Discussion Paper*, Ottawa: Genome Canada, November.

Giersch, H. (ed.) (1982). *Emerging Technologies: Consequences for Economic Growth, Structural Change, and Employment*, Institut für Weltwirtschaft an der Universität Kiel.

Gofton, L. and Haimes, E. (1999). "Necessary evils? Opening up in sociology and biotechnology", *Sociological Research Online*, <http://www.socresonline.org.uk/socresonline/-1995/1/gofton.html>

Gould, C. C. (1993). "Contemporary Legal Conceptions of Property and Their Implications for Democracy", in Gould, C. C. *Rethinking Democracy: Freedom and Social Cooperation in Politics, Economy and Society*, Cambridge: Cambridge University Press.

Government of Canada (1999). *Canadian Biotechnology Statistics*, Ottawa.

Government of Canada (1999). *Canadian Environmental Protection Act*, Canada Gazette Part II, November 24 edition, Ottawa: Government of Canada.

Grace, E. (1997). *Biotechnology Unzipped*, Washington: National Academy Press.

Granberg, A. (1997). "A Mapping the Cognitive and Institutional Structures of an Evolving Advanced-Materials Field: The Case of Powder Technology", in Carlsson, B. (ed.) *Technological Systems and Industrial Dynamics*, Boston, Dordrecht and London: Kluwer Academic Publishers.

Granstrand, O. and Jacobsson, S. (1991). "A When are Technological Changes Disruptive? - A Preliminary Analysis of Intervening Variables Between Technological and Economic Changes", paper presented at the Marstrand Symposium on Economics of Technology, Marstrand, August.

Grant, R. (1991). "The resource based theory of competitive advantage: implications for strategy formulation", *California Management Review*, Vol. 33, No. 3, pp. 114-135.

Greenwood, R. and Hinings, C. R. (1993). "Understanding Strategic Change: the contribution of Archetype", *Academy of Management Journal*, Vol. 36, pp. 1052-1081.

Greis, N. P., Dibner, M. D. and Bean, A. S. (1994). "External partnering as a response to innovation barriers and global competition in biotechnology", *Research Policy*, Vol. 24.

Griliches, Z. (1990). "Patent Statistics as Economic Indicators: A Survey", *Journal of Economic Literature*, Vol. 28, No. 3, pp. 1661-1707.

Griliches, Z. (1992). "The Search for R&D Spillovers", *Scandinavian Journal Of Economics*, Vol. 94, Supplement 29-47.

Griliches, Z. (1994). "Productivity, R&D, and the Data Constraint", *American Economic Review*, Vol. 84, pp. 1-22.

Grupp, H. (1996). "A Spillover Effects and the Science Base of Innovations Reconsidered: An Empirical Approach", *Journal of Evolutionary Economics*, Vol. 6, pp. 175-197.

Habermas, J. (1981). *Theorie des kommunikativen Handelns*, 2 Volumes. Frankfurt am Main: Suhrkamp.

Habermas, J. (1987). *Die neue Unübersichtlichkeit*, Frankfurt am Main: Suhrkamp.

Hakansson, P., Kjellberg, H. and Lundgren, A. (1993). "Strategic Alliances in Global Biotechnology – A Network Approach", *International Business Review*, Vol. 2, No. 1, pp. 65-82.

Hamel, G. (1991). "Competition for Competence and Interpartner Learning within International Strategic Alliances", *Strategic Management Journal*, Vol. 12, pp. 83-103.

Harris, R. A. and Milkis, S. M. (1989). *The Politics of Regulatory Change*, New York: Oxford University Press.

Hellman, J.-L. (1993). *Forskningsbaserat medicinst-tekniskt företagande*, B 1993:12, Stockholm: NUTEK.

Henderson, R., Orsenigo, L. and Pisano, G. P. (1999). "The Pharmaceutical Industry and the Revolution in Molecular Biology: Interactions Among Scientific, Institutional and Organizational Change", in Mowery, D. C. and Nelson, R. R. (eds) *Sources of Industrial Leadership*, Cambridge: Cambridge University Press.

Henrekson, M. and Johansson, D. (2000). "Entreprenörskapet som tillväxtmotor i EU", in Bernitz, Gustavsson, Oxelheim, (eds) *Europaperspektiv*, Yearbook 2000, Santerus Förlag.

Hill, C. (1995). "Diversification and economic performance: Bringing structure and corporate management back into the picture", in Rumelt, R. P., Schendel, D. E. and Teece, D. J. (eds) *Fundamental Issues in Strategy*, Boston: Harvard Business School Press.

Hirschman, A. O. (1977). *The Passions and the Interests*, Princeton, New Jersey: Princeton University Press.

Hodgson, J. (2000). "Crystal gazing the new technologies", *Nature Biotechnology*, Vol. 18, No. 1, pp. 29-31.

Holmén, M. (1998). "Regional Industrial Renewal: the Growth of Microwave Technology in West Sweden", mimeo, Department of Industrial Dynamics, School of Technology Management and Economics, Chalmers University of Technology, Göteborg, Sweden.

Holmén, M. and Jacobsson, S. (1998). "A method for identifying actors in a knowledge based cluster", forthcoming in *Economics of Innovation and New Technology*.

Houdebine, L. (1996). "Animal Biotechnology", *STI Review*, No. 19, 75-92.

House of Lords (1998). *EC Regulation of Genetic Modification in Agriculture*, Report, Select Committee on the European Communities, London: HMSO.
http://europa.eu.int/comm/dg24/health/sc/scv/out19_enhtml#_Toc446386014.

Hughes, T. (1987). "A The Evolution of Large Technological Systems", in Bijker, W. E., Hughes, T. P. and Pinch, T. J. (eds) *The Social Construction of Technological Systems*, Cambridge, MA and London: MIT Press.

Hunt, W. A. (1999). "Genetically-Modified Politics", *Policy Options*, November, pp. 59-62.

Intersociety Working Group (1997). "AAAS Report XXII Research and Development FY 1998", Washington DC: American Association for the Advancement of Science.

Joly, P. B. and de Looze, M. A. (1996). "An analysis of innovation strategies and industrial differentiation through patent applications: the case of plant biotechnology", *Research Policy*, Vol. 25, pp.1027-1046.

Jordan, A. and O'Riordon, T. (1995). "The Precautionary Principle in UK Environmental Policy Making", in Gray, T. S. (ed.) *UK Environmental Policy in the 1990s*, London: MacMillan.

Kemp, R. P. M., Mulder, P. and Reschke, C. H. (1999). "Evolutionary Theorising on Technological Change and Sustainable Development", OCFEB Working Paper 9912, Research Centre for Economic Policy, Erasmus University Rotterdam.

Kenney, M. (1986). *Biotechnology. The University-Industry Complex*, New Haven and London: Yale University Press.

Knight, B. E. A. and Whitehead, R. (1994). "Transgenic Crops", AGROW, PJB Publications Ltd.

Knight, F. (1921). *Risk, Uncertainty and Profit*, Boston: Houghton-Mifflin.

Köhler, G. and Milstein, C. (1975). "Continuous Cultures of Fused Cells Secreting Antibody of Predefined Specificity", *Nature*, Vol. 256, pp. 495-497.

Krasner, D. (ed.) (1983). *International Regimes*, Ithaca: Cornell University Press.

Laestadius, S. (1998). "The relevance of science and technology indicators: the case of pulp and paper", *Research Policy*, Vol. 27, pp.385-395.

Laestadius, S. (2000). A Prospective Paradigm Shift in Forest Industry. Mimeo, KTH, Indek.

Lal Das, B. (1999). *The World Trade Organization: A Guide to The Framework For International Trade*, London: Zed Books.

Larson, A., Lembre, P. and Meldahl, C. (1998). "Science Parks and Industrial Development – a competence bloc analysis of Swedish, Taiwanian and US Industrial Districts", short version of Masters thesis, KTH, Stockholm.

Lebow, I. (1995). "Information Highways and Byways", New York: IEEE Press.

Lederer, E. (1992). *Grundzüge der ökonomischen Theorie*, Eine Einführung. Tübingen. J.C. B. Mohr (Paul Siebeck).

Lee, K. and Burrill, G. S. (1996). "Biotech 97 Alignment: The Eleventh Industry Annual Report", Palo Alto, CA: Ernst and Young LLP.

Lerner, J. and Merges R. P. (1998). "The control of technology alliances: An empirical analysis of the biotechnology industry", *Journal of Industrial Economics*, Vol. 46, pp.125-156.

Liebeskind, J. P., Oliver, A. L., Zucker, L. and Brewer, M. B. (1995). "Social Networks, Learning and flexibility: Sourcing Scientific Knowledge in New Biotechnology Firms", Working Paper W5320, Cambridge MA: National Bureau of Economic Research.

Liebeskind, J. P., et al. (1996). "Social Networks, Learning and Flexibility: Sourcing Scientific Knowledge in Biotechnology Firms", *Organization Science*, Vol. 7, No. 4, pp. 428-443.

Lindh, T. (1993). "Lessons from Learning to have Rational Expectations", in Day, R. H., Eliasson, G., and Wihlborg, C. G. (eds) *The Markets for Innovation, Ownership and Control*, Stockholm, North-Holland, Amsterdam, London, New York, Tokyo: IUI.

Luhmann, N. (1988). *Die Wirtschaft der Gesellschaft*, Frankfurt am Main: Suhrkamp.

Lundvall, B. -Å. (1988). "Innovation as an Interactive Process: From User-Supplier Interaction to the National System of Innovation", in Dosi, G. et al. (eds) *Technical Change and Economic Theory*, London: Pinter.

Lundvall, B.-Å. (ed.) (1992). *National Systems of Innovation: Towards a Theory of Innovation and Interactive Learning*, London: Pinter.

Lynas, M. (1999). "The World Trade Organization and GMOs", *Consumer Policy Review*, Vol. 9, No. 6, Nov.-Dec., pp. 214-219.

MacDonald, M. (2000). "Socio-economic Versus Science-Based Regulation: Informal Influences on the Formal Regulation of rbST in Canada", in Doern, G. B. and Reed, T. (eds) *Risky Business: Canada's Changing Science-Based Policy and Regulatory Regime*, Toronto: University of Toronto Press.

Mahoney, J. and Pandrian, J. R. (1992). "The resource-based view within the conversion of strategic management", *Strategic Management Journal*, Vol. 13, pp. 363-380.

Malerba, F. and Orsenigo, L. (1990). "Technological Regimes and Patterns of Innovation: A Theoretical and Empirical Investigation of the Italian Case", in Heertje, A. and Perlman, M. (eds) *Evolving Technology and Market Structure*, Ann Arbor: Michigan University Press.

Malerba, F. and Orsenigo, L. (1993). "Technological Regimes and Firm Behavior", *Industrial and Corporate Change*, Vol. 2, No. 1, pp. 45-71.

Malerba, F. and Orsenigo, L. (1995). "Schumpeterian Patterns of Innovation", *Cambridge Journal of Economics*, Vol. 19, No. 1, pp. 47-65.

Malinvaud, E. (1967). "Decentralized Procedures in Planning", in Malinvaud, E. and Bacharach, M. O. L. (eds) Activity *Analysis in the Theory of Growth and Planning*, London: Macmillan.

Mansfield, E. (1994). "Academic Research Underlying Industrial Innovations: Sources, Characteristics and Financing", *Review of Economics and Statistics*, Vol. 77, pp. 55-62.

Marshall, A. (1890). *Principles of Economics*, New York and London: Macmillan & Co.

Marshall, A. (1919). *Industry and Trade*, London: Macmillan & Co.

Martin, J. L. (1999). "The myth of the consumption-oriented economy and the rise of the desiring subject", *Theory and Society*, Vol. 28, pp. 425-453.

Mazur, A. ([1989] 1999). "Connections: biomedical sciences in supermarket tabloids", *Knowledge, Technology, & Policy*, Vol. 12, pp. 19-26.

McKelvey, M. (1993). "Technologies embedded in nations? Genetic Engineering and technological change in national systems of innovation", *The Journal of Socio-Economics*, Vol. 22, No. 4, pp. 353-377.

McKelvey, M. (1996). *Evolutionary Innovation. The Business of Biotechnology*, New York: Oxford University Press.

McLean, I. (1987). *Public Choice. An Introduction*, Oxford: Basil Blackwell.

McMillan, G. S., Narin, F. and Deeds, D. L. (2000). "An analysis of the critical role of public science in innovation: the case of biotechnology", *Research Policy*, Vol. 29, No. 1, pp. 1-8.

Mihill, C. (1998). "Drug firm donates £1 bn to defeat tropical disease", *Guardian*, 27 January, 9.

Milward, A. S. (1992). *The European Rescue of the Nation-State*, Berkeley: University of California Press.

Mironesco, C. (1998). "Parliamentary Technology Assessment of Biotechnologies: A Review of Major TA Reports in the European Union and the USA", *Science and Public Policy*, Vol. 24, No. 5, (October), pp. 327-342.

Miyazaki, K. (1994). "A Interlinkage Between Systems, Key Components and Component Generic Technologies in Building Competencies", *Technology Analysis & Strategic Management*, Vol. 6, No. 1.

Moffat, A. S. (1998). "Improving gene transfer into livestock", *Science*, Vol. 282, pp. 1619-1620.

Momma, S. and Sharp, M. (1999). "Developments in new biotechnology firms in Germany", *Technovation*, Vol. 19, pp. 267-282.

Monsan, P. (1999). "Vingt ans de biotechnologie en France", *Biofutur*, Vol. 194, pp. 23-27.

Morrison, S. W. and Giovannetti, G. T. (1999). Bridging the Gap, 13th Biotechnology Industry Annual Report, Palo Alto, Calif.: Ernst & Young, LLP.

Münch, R. (1990). "Differentiation, rationalization, interpenetration: the emergence of modern society", in Alexander, J. C. and Colomy, P. (eds) *Differentiation Theory and Social Change*, New York: Columbia University Press.

Münch, R. (1991). *Die Dialektik der Kommunikationsgesellschaft*, Frankfurt am Main: Suhrkamp.

Münch, R. (1992). "The dynamics of societal communication", in Colomy, P. (ed.), *The Dynamics of Social Systems*, London: Sage.

Nadiri, I. (1978). "A Dynamic Model of Research and Development Expenditure", in Carlsson, B., Eliasson, G. and Nadiri, I. (eds) *The Importance of Technology and the Permanence of Structure in Industrial Growth,* Stockholm: IUI Conference Reports, 1978:2.

Nadiri, I. (1993). "Innovations and Technological Spillovers", Working Paper No. 4423. Cambridge, MA: National Economic Research Bureau.

Nelkin, D. (1995). "The power of DNA diagnostics", Genethics, Basel: Ciba-Geigy Ltd.

Nelson, R. R. (1988a). "National Systems of Innovation: Preface", in Dosi, G. et al., (eds) *Technical Change and Economic Theory*. London: Pinter.

Nelson, R. R. (1988b). "Institutions Supporting Technical Change in the United States", in Dosi, G. et al., (eds) *Technical Change and Economic Theory*. London: Pinter.

Nelson, R. R. (1995). "Why do firms differ and how does it matter?", in Rumelt, R. P., Schendel, D.E. and Teece, D. J. (eds.) *Fundamental Issues in Strategy*, Boston: Harvard Business School Press.

Nelson, R. R. (ed.) (1993). *National Innovation Systems: A Comparative Study*, New York: Oxford University Press.

Nelson, R. R. and Winter, S. (1977). "In Search of a Useful Theory of Innovation", *Research Policy*, Vol. 6, No. 2, pp. 36-76.

Nelson, R. R. and Winter, S. (1982). *An Evolutionary Theory of Economic Change*, Cambridge, MA: Harvard University Press.

Nightingale, P. (2000). "Economies of Scale in Experimentation: Knowledge and Technology in Pharmaceutical R&D", *Industrial and Corporate Change*.

Niosi, J. (1995). *Flexible Innovation. Technological Alliances in Canadian Industry*, Montreal and Kingston: McGill-Queen's University Press.

Niosi, J. (2000). "Science-based industries, a new Schumpeterian taxonomy", *Technology in Society*, (in print).

Niosi, J. and Bas, T. G. (2000). "The Competencies of Regions. Canada's Clusters in Biotechnology", *International Journal of Biotechnology*, Vol. 2, No. 4, (forthcoming).

Niosi, J., Saviotti, P. P., Bellon, B. and Crow, M. (1993). "National Systems of Innovation: In Search of a Workable Concept", *Technology in Society*, Vol. 15, No. 2, pp. 207-227.

O'Mahony, P. (ed.) (1999). *Nature, Risk and Responsibility. Discourses of Biotechnology*, New York: Routledge.

O'Riordan, T. (1996). "Exploring the Role of Civic Science in Risk Management", in Hood, C. and Jones, D. (eds) (1996). *Accident and Design: Contemporary Debates in Risk Management*, London: UCL Press.

OECD (1982). *Biotechnology: International Trends and Perspectives*, Paris: OECD.

OECD (1993). *Proposed Standard Practice for Surveys of Research and Experimental Development - The Frascati Manual*, Paris: OECD.

OECD (1994). *Using Patent Data as Science and Technology Indicators-Patent Manual*, Paris: OECD.

OECD (1995). *Literacy, Economy and Society*, Paris: OECD.

OECD (1995b). *The Measurement of Human Resources Devoted to S &T-Canberra Manual*, OECD/GD (94) 114, Paris: OECD.

OECD (1997). *Proposed Guidelines for Collecting and Interpreting Technological Innovation Data - The Oslo Manual*, Paris: OECD.

OECD (1998). *Biotechnology for Clean Industrial Products and Processes*, Paris: OECD.

OECD (1999a). *Science, Technology and Industry Scoreboard 1999, Benchmarking Knowledge-Based Economies*, Paris: OECD.

OECD (1999b). Biotech Database, http://www.olis.oecd.org/bioprod.nsf.

OECD (1999c). Biotech Database, http://www.oecd.org/ehs/summary.htm.

OECD/Eurostat (1997). *Proposed Guidelines for Collecting and Interpreting Technological Innovation Data- Oslo Manual*, Paris/Luxembourg: OECD/Eurostat.

Office of Technology Assessment (OTA) (1991). "Biotechnology in a Global Economy", Office of Technology Assessment, Washington DC: Congress of the United States.

Oliver, R. W. (2000). *The Coming Biotech Age*, New York: McGraw-Hill.

Orsenigo, L. (1989). *The Emergence of Biotechnology*, London: Pinter.

Paradis, K., Langford, G., Long, Z., Heneine, W., Sandstrom, P., Switzer, W., Chapman, L., Lockey, C., Onions, D., The XEN III Study Group and Otto, E. (1999). "Search for cross-species transmission of porcine endogenous retrovirus in patients treated with living pig tissue", *Science*, Vol. 285, No. 5431, pp. 1236-1241.

Pattinson, W., Van Beuzekom, B. and Wyckoff, A. (2000). "Internationally Comparable Indicators on Biotechnology: A Stocktaking, A Proposal for Work, and Supporting Material", Working Paper, Science, Innovation and

Electronic Information Division, Ottawa: Statistics Canada. (available at: www.statcan.ca/english/research.scilist.htm)

Pavitt, K. (1988). "Uses and Abuses of Patent Statistics", in Van Raan, (ed.) *Handbook of Quantitative Studies of Science and Technology*, Amsterdam: Elsevier Science Publishers, B.V.

Peacock, A. (1984). *The Regulation Game*, Oxford: Basil Blackwell.

Penrose, E. (ed.) (1959). "The Theory of the Growth of the Firm", Oxford: Basic Blackwell.

Peteraf, M. (1993). "The cornerstones of the competitive advantage: A resource based view", *Strategic Management Journal*, Vol. 14, pp. 179-191.

Petersman, E. U. and Marceau, G. (1997). "The GATT/WTO Dispute Settlement System International Law, International Organization and Dispute Settlement", *Journal of World Trade*, Vol. 31, No. 3, pp. 169-179.

Pfeffer, J. and Salancik, G. R. (1978). *The external control of organizations: a resource dependence perspective*, New York: Harper & Row Publishers.

Pisano, G. (1988). "Innovation through markets, hierarchies, and joint ventures: Technology strategy and collaborative arrangements in the biotechnology industry", Doctoral dissertation, UC Berkeley.

Pisano, G. (1991). "The governance of innovation: vertical integration and collaborative arrangements in the biotechnology industry", *Research Policy*, Vol. 20, pp. 237-249.

Porter, M. (1980). *Competitive strategy*, New York: Free Press.

Porter, M. (1990). *The Competitive Advantage of Nations*, New York: The Free Press.

Porter, M. (1998). "A Clusters and the New Economics of Competition", *Harvard Business Review*, Nov.-Dec., pp. 77-90.

Powell, W. W. (1990). "Neither market nor hierarchy: networks forms of organization Research", *Organizational Behavior*, Vol. 12, pp. 295-336.

Powell, W. W. (1998). "Learning form Collaboration: Knowledge and Networks in the Biotechnology and Pharmaceutical Industries", *California Management Review*, Vol. 40, No. 3, pp. 228-240.

Powell, W. W. (1999). "The social construction of an organizational field: the case of biotechnology", *International Journal of Biotechnology*, Vol. 1, No. 1, pp. 42-66.

Powell, W. W., Koput, K. W. and Smith-Doerr, V (1996). "Inter-organizational collaboration and the locus of innovation: networks of learning in biotechnology", *Administrative Science Quarterly*, Vol. 41, pp. 116-145.

Praest, M. (1998). "A Patterns of Technological Competence Accumulation and Performance: Experiences from Telecommunications", Ph.D. thesis, Aalborg University, October.

Prakash, C. S. (1999). "Feeding a world of six billion", *AgbioForum*, Vol. 2, pp. 223-225.

Prevezer, M. (1998). "Clustering in biotechnology in the USA", in Swan, G. M. P., Prevezer, M. and Stout, D. (ed.) *The Dynamics of Industrial Clustering*, New York: Oxford University Press.

Purdue, D. (1995). "Hegemonic Trips: World Trade, Intellectual Property and Biodiversity", *Environmental Politics*, Vol. 4, No. 1 (Spring), pp. 88-107.

Ranson, S., Hinings, C. R. and Greenwood, R. (1980). "The Structuring of Organizational Structures", *Administrative Science Quarterly*, Vol. 25, pp. 1-17.

Rapoport, A. (1998). How Has the Field Mix of Academic R&D Changed? Issue Brief, Arlington, Va.: National Science Foundation, Division of Science Resources Studies.

Rappa, M. (1994). "Assessing the Rate of Technological Progress Using Hazard Rate Models of R&D Communities", *R&D Management*, Vol. 24, No. 2.

Rehm, H. J. (1982). "The Economic Potential of Biotechnology", in Giersch, H. (ed.) *Emerging Technologies: Consequences for Economic Growth, Structural Change, and Employment*, Institut für Weltwirtschaft an der Universität Kiel.

Reid, M. (2000). "Buying Into The Gene Pool", *Shares*, March 16, pp. 24-27.

Rickne, A. (2000). "The Growth of New Technology-Based Firms- The Case of Biomaterials in Sweden, Massachusetts and Ohio", in Carlsson, B. (ed.) *New Technological Systems in the Bio Industries: An International Study*, Boston, Dordrecht and London: Kluwer Academic Publishers, (forthcoming).

Rifkin, J. (1998). *The Biotechnology Century*, New York: Tarcher/Putnam Books.

Romer, P. M., (1986). "Increasing Returns and Long-Run Growth", *Journal of Political Economy*, Vol. 94, No. 5, Oct., pp. 1002-1037.

Romer, P. M. (1990). "Endogenous Technological Change", *Journal of Political Economy*, Vol. 98, No. 5, pt. 2, pp. 71-102.

Ronchi, E. (1996). "Biotechnology and the new revolution in health care and pharmaceuticals", *STI Review*, No. 19, pp. 19-43.

Rose A., (1998). *Biotechnology Use by Canadian Industry - 1996*, Ottawa, Statistics Canada, Working Paper, Science, Innovation and Electronic Information Division, Ottawa: Statistics Canada.

Rose, N. (1999). *Powers of Freedom. Reframing Political Thought*, Cambridge: Cambridge University Press.

Russo, M. V., and Fouts, P. A. (1997). "A resource-based perspective on corporate environmental performance and profitability", *Academy of Management Journal*, Vol. 40, No. 3, pp. 534-559.

Ruttan, V. W. (1999). "Biotechnology and agriculture: A skeptical perspective", *AgBioForum*, Vol. 2, pp. 54-60.

Ryan, A., Freeman, J. and Hybels, R. (1995). "Biotechnology Firms", in Carroll, G. R. Hannan, M. T. (eds) *Organizations in Industry*, New York: Oxford University Press.

Sagar, A., Daemmrich, A. and Ashiya, M. (2000). "The tragedy of the commoners: biotechnology and its publics", *Nature Biotechnology*, Vol. 18, No. 1, pp. 2-4.

Salter, W. E. G. (1960). *Productivity and Technical Change*, Cambridge MA: Cambridge University Press.

Saxenian, A. (1994). *Regional Advantage. Culture and Competition in Silicon Valley and Route 128*, Cambridge, MA: Harvard University Press.

Schumpeter, J. A. (1911) (English edition 1934). *The Theory of Economic Development*, Harvard Economic Studies. Vol. XLVI. Cambridge, MA.: Harvard University Press.

Schumpeter, J. A. (1942). *Capitalism, Socialism and Democracy*, New York: Harper & Row.

Schwartz, B. (1999). "Capitalism, the market, the 'underclass,' and the future", *Society*, Vol. 37, pp. 33-42.

Science Council of Canada (1989). *Enabling Technologies; Springboard for a Competitive Future*, Ottawa: Science Council of Canada.

Scientific Committee on Veterinary Measures Relating to Public Health (1999). "Report on Public Health Aspects of the Use of Bovine Somatotrophin", EC DG24, Brussels available November 1999 on

Sell, S. K. (1995). "Intellectual Property Protection and Antitrust in the Developing World: Crisis, Coercion and Choice", *International Organization*, Vol. 49, No. 2, pp. 315-349.

Sell, S. K. (1998). *Power and Ideas: North-South Politics of Intellectual Property and Antitrust*, New York: State University of New York Press.

Senker, J. (1996). "National systems of innovation, organizational learning and industrial biotechnology", *Technovation*, Vol. 16, No. 5, pp. 219-230.

Senker, J., Joly, P. B. and Reinhard, M. (1996). "Overseas Biotechnology Research by Europe's Chemical/Pharmaceutical Multinationals: Rationale and Implications", SPRU, STEEP Discussion Paper, No. 33.

Shan, W. (1996). "High-Tech Entrepreneurship and Organizational Choice", in Niosi, J. (ed.) *New Technology Policy and Social Innovations in the Firm*, London: Pinter.

Sharp, M. (1996). *The Science of Nations: European Multinationals and American Biotechnology*, SPRU, STEEP Discussion Paper N. 28, 40 p.

Sharp, M. and Galimberti, I. (1994). "Co-operative alliances and internal competences: some case studies in biotechnology", SPRU, University of Sussex.

Shiva, V. (1997). *Biopiracy: The Plunder of Nature and Knowledge*, Boston: South End Press.

Shohet, S. (1998). "Clustering and UK biotechnology", Swan, G. M. P., Prevezer, M. and Stout, D. (eds) *The Dynamics of Industrial Clustering*, New York: Oxford University Press, pp. 194-224.

Shrivastava, P., Huff, A. S. and Dutton, J. E. E. (1994). "Resource-based views of the firm. Advances in Strategic Management", vol. Ed.^Eds. 10A. Greenwich, Conn. and London: JAI Press.**

Simonin, B. L. (1997). "The Importance of Collaborative Know-how: an Empirical test of the Learning Organization", *Academy of Management Journal*, Vol. 40, No. 5, pp. 1150-1174.

Smith, W. and Halliwell, J. (1999). *Principles and Practices For Using Scientific Advice in Governmental Decision Making: International Best Practices*, Report to the S&T Strategy Directorate, Ottawa: Industry Canada.

Sombart, W. ([1916] 1921). *Der moderne Kapitalismus. Historisch-systematische Darstellung des gesamten Wirtschaftslebens von seinen Anfaengen bis zur Gegenwart. Erster Band: Einleitung- Die vorkapitalistische Wirtschaft — Die historischen Grundlagen des modernen Kapitalismus*. Erster Halbband. München und Leipzig: Duncker & Humblot.

Statistics Canada (1980). *Standard Industrial Classification - 1980*, Catalogue. No. 12-501E, Ottawa: Statistics Canada.

Statistics Canada (1997). *Biotechnology Research and Development in Canadian Industry, 1995*, Catalogue No. 88-001-XPB, Vol. 21 No. 11, Ottawa: Statistics Canada.

Statistics Canada (1998a). *Biotechnology Scientific Activities in Selected Federal Government Departments and Agencies, 1997-98*, Catalogue No. 88-001-XPB, Vol. 22, No. 4, Ottawa: Statistics Canada.

Statistics Canada (1998b). *Science and Technology Activities and Impacts: A Framework for a Statistical Information System*, Catalogue No. 88-522-XPB, Ottawa: Statistics Canada.

Stehr, N. (1999). "The future of inequality", *Society*, Vol. 36, pp. 54-59.

Stehr, N. (2000). *Knowledge and Economic Conduct: The Social Foundations of the Modern Economy*, Toronto: University of Toronto Press.

Stigler, G. (1971). "The theory of economic regulation", *Bell Journal of Economics and Management*, Vol. 2, pp. 3-21.

Stirling, A. (1999). *On Science and Precaution in the Management of Technological Risk*, Science Policy Research Unit, University of Sussex.

Straughan, R. (1995). "Monsters and morality", *Genethics*, Ciba-Geigy Ltd. Basel.

Swan, G. M. P., Prevezer, M. and Stout, D. (1998). *The Dynamics of Industrial Clustering. International Comparisons in Computing and Biotechnology*, New York: Oxford University Press.

Sylos-Labini, P. (1969). *Oligopoly and Technical Progress*, Cambridge, MA: Harvard University Press.

Teece, D. J. (1986). "Profiting from Technological Innovation: Implications for Integration, collaboration, licensing and Public Policy", *Research Policy*, Vol. 15, pp. 285-305.

Teece, D. J. (1996). "Firm Organization, Industry Structure and Technological Innovation", *Journal of Economic Behavior and Organization*, Vol. 31, pp. 193-224.

The Economist (1999). "Genomic Pronouncements", *The Economist*, December 4, pp. 77-78.

Thurow, L. (1999). *Building Wealth*, Boston: MIT Press.

Tolbert, P. S. (1985). "Resource dependence and institutional environments: sources of administrative structure in institutions of higher education", *Administrative Science Quarterly*, Vol. 30, pp. 1-13.

Touraine, A. ([1992] 1995). *The Critique of Modernity*, Oxford: Blackwell.

Trebilcock, M. and Howse, R. (1995). *The Regulation of International Trade*, London: Routledge.

Turney, J. (1998). *Frankenstein's Footsteps. Science, Genetics and Popular Culture*, New Haven: Yale University Press.

U.S. Embassy (2000). Byliner: Assistant Secretary of State Sandalow on Biosafety Protocol, February 11, 2000.

UNEP (2000). *Draft Cartagena Protocol on Biosafety*, Montreal: UNEP.

US Patent and Trademark Office. *US Patent Database*.

Valette, A. (1994). "La formation des trajectoires d'offre de soins : les interactions hopital-environnement ", Paris : CRG-Université Paris IX Dauphine.

Van den Daele, W. (1992). "Concepts of nature in modern societies and nature as a theme in sociology", in Dierkes, M. and Biervert, B. (eds) *European Social Science in Transition: Assessment and Outlook*, Frankfurt am Main: Campus.

Van Dijick, J. (1998). *Imagenation. Popular Images of Genetics*, New York: New York University Press.

Vogel, D. (1995). *Trading Up: Consumer and Environmental Regulation in a Global Economy*, Cambridge, MA: Harvard University Press.

Vogel, D. (1998). "The Globalization of Pharmaceutical Regulation", *Governance*, Vol. 11, No. 1, pp. 1-22.

Walsh, V., Niosi, J. and Mustar, P. (1995) "Small-firm formation in biotechnology: a comparison of France, Britain and Canada", *Technovation*, Vol. 15, No. 5, pp. 303-327.

Watson, J. D. and Crick, F.H.C. (1953). "Molecular Structure of Nucleic Acids", *Nature*, Vol. 171, pp.737-738.

Weber, M. ([1922] 1978). *Economy and Society*, [Roth, G. and Wittich, C. (eds.),] Berkeley, California: University of California Press.

Webster, F. and Robins, K. (1989). "Plan and control: towards a cultural history of the information society", *Theory and Society*, Vol. 18, pp. 323-351.

Weick, K. (ed.) (1979). *A Social Psychology of Organizing*, Reading, MA: Addison Westley.

Wellcome Trust (1999). "From Genome to Health", *Wellcome News*, Issue 20, Q 3.

Wernerfelt, B. (1984). "A resource-based view of the firm", *Strategic Management Journal*, Vol. 5, pp. 171-180.

Wilmut, I., Campbell, K. and Tudge, C. (1999). *The Second Creation*, London: Headline.

Wilson, G. K. (1980). *Interest Groups*, Oxford: Basil Blackwell.

Ye, X. D., AlBabili, S., Kloti, A., Zhang, J., Lucca, P., Beyer, P. and Potrykus, I. (2000). "Engineering the provitamin A (beta-carotene) biosynthetic pathway into (carotenoid-free) rice endosperm", *Science*, Vol. 287, No. 5451, pp. 303-305.

Ylä-Anttila, P. (1994). "An Industrial Clusters - a Key to New Industrialisation?", *Kansalis Economic Review*, 1/1994, pp. 4-11.

Zacher, M. W. (with Brent A. Sutton) (1996). *Governing Global Networks: International Regimes For Transportation and Communications*, New York: Cambridge University Press.

Zucker, L. G. and Darby, M. R., (1998). *Present at the Revolution: Transformation of Technical Identity for a Large Incumbent Pharmaceutical Firm After the Biotechnological Breakthrough*, Working Paper W5243, Cambridge, MA: National Bureau of Economic Research.

Zucker, L. G. and Darby, M. R. (1999). "Star-scientist linkages to firms in APEC and European countries: indicators of regional institutional differences affecting competitive advantage", *International Journal of Biotechnology*, Vol. 1, No. 1, pp. 119-131.

Zucker, L. G., Darby, M. R. and Brewer, M. (1998). "Intellectual Human Capital and the Birth of U.S. Biotechnology Enterprises", *American Economic Review*, Vol. 88, No. 1, pp. 290-306.

Zucker, L., Darby, M. R. and Armstrong, J. (1994). "Intellectual Capital and the Firm: The Technology of Geographically Localized Knowledge Spillovers", Working Paper, Cambridge MA: National Bureau of Economic Research.

Zucker, L. G., Darby, M. R., and Armstrong, J. (1998a). "Geographically Localized Knowledge: Spillovers or Markets", *Economic Inquiry*, Vol. 36, No. 1, pp. 65-86.

Zucker, L. G., Darby, M. R. and Brewer, M. B. (1998b). "Intellectual human capital and the birth of US biotechnology enterprise", *American Economic Review*, March (88) 1, pp. 290-306.

INDEX

Economics of Science, Technology and Innovation

1. A. Phillips, A.P. Phillips and T.R. Phillips:
   Biz Jets. Technology and Market Structure in the Corporate Jet Aircraft Industry. 1994 — ISBN 0-7923-2660-1
2. M.P. Feldman:
   The Geography of Innovation. 1994 — ISBN 0-7923-2698-9
3. C. Antonelli:
   The Economics of Localized Technological Change and Industrial Dynamics. 1995 — ISBN 0-7923-2910-4
4. G. Becher and S. Kuhlmann (eds.):
   Evaluation of Technology Policy Programmes in Germany. 1995 — ISBN 0-7923-3115-X
5. B. Carlsson (ed.): *Technological Systems and Economic Performance: The Case of Factory Automation.* 1995 — ISBN 0-7923-3512-0
6. G.E. Flueckiger: *Control, Information, and Technological Change.* 1995 — ISBN 0-7923-3667-4
7. M. Teubal, D. Foray, M. Justman and E. Zuscovitch (eds.):
   Technological Infrastructure Policy. An International Perspective. 1996 — ISBN 0-7923-3835-9
8. G. Eliasson:
   Firm Objectives, Controls and Organization. The Use of Information and the Transfer of Knowledge within the Firm. 1996 — ISBN 0-7923-3870-7
9. X. Vence-Deza and J.S. Metcalfe (eds.):
   Wealth from Diversity. Innovation, Structural Change and Finance for Regional Development in Europe. 1996 — ISBN 0-7923-4115-5
10. B. Carlsson (ed.):
    Technological Systems and Industrial Dynamics. 1997 — ISBN 0-7923-9940-4
11. N.S. Vonortas:
    Cooperation in Research and Development. 1997 — ISBN 0-7923-8042-8
12. P. Braunerhjelm and K. Ekholm (eds.):
    The Geography of Multinational Firms. 1998 — ISBN 0-7923-8133-5
13. A. Varga:
    University Research and Regional Innovation: A Spatial Econometric Analysis of Academic Technology Transfers. 1998 — ISBN 0-7923-8248-X
14. J. de la Mothe and G. Paquet (eds.):
    Local and Regional Systems of Innovation — ISBN 0-7923-8287-0
15. D. Gerbarg (ed.):
    The Economics, Technology and Content of Digital T V — ISBN 0-7923-8325-7
16. C. Edquist, L. Hommen and L. Tsipouri
    Public Technology Procurement and Innovation — ISBN 0-7923-8685-X
17. J. de la Mothe and G. Paquet (eds.):
    Information, Innovation and Impacts — ISBN 0-7923-8692-2

18. J. S. Metcalfe and I. Miles (eds.):
Innovation Systems in the Service Economy: Measurement and Case Study Analysis ISBN 0-7923-7730-3
19. R. Svensson:
Success Strategies and Knowledge Transfer in Cross-Border Consulting Operations ISBN 0-7923-7776-1
20. P. Braunerhjelm:
Knowledge Capital and the "New Economy": Firm Size, Performance and Network Production ISBN 0-7923-7801-6
21. J. de la Mothe and J. Niosi (eds):
The Economic and Social Dynamics of Biotechnology ISBN 0-7923-7922-5

KLUWER ACADEMIC PUBLISHERS — BOSTON / DORDRECHT / LONDON